Palgrave Studies in Medieval and Early Modern Medicine

Series Editors

Jonathan Barry
Department of History
University of Exeter
Exeter, UK

Fabrizio Bigotti
Institute for the History of Medicine
Julius Maximilian University
Würzburg, Germany

CENTRE FOR THE STUDY OF
MEDICINE AND THE BODY
IN THE RENAISSANCE

The series focuses on the intellectual tradition of western medicine as related to the philosophies, institutions, practices, and technologies that developed throughout the medieval and early modern period (500-1800). Partnered with the Centre for the Study of Medicine and the Body in the Renaissance (CSMBR), it seeks to explore the range of interactions between various conceptualisations of the body, including their import for the arts (e.g. literature, painting, music, dance, and architecture) and the way different medical traditions overlapped and borrowed from each other. The series particularly welcomes contributions from young authors. The editors will consider proposals for single monographs, as well as edited collections and translations/editions of texts, either at a standard length (70-120,000 words) or as Palgrave Pivots (up to 50,000 words).

Matteo Pace
Editor

Dante and the Sciences of the Human

Medicine, Physics, and the Soul

palgrave
macmillan

Editor
Matteo Pace
Department of Italian Studies
Connecticut College
New London, CT, USA

ISSN 2524-7387 ISSN 2524-7395 (electronic)
Palgrave Studies in Medieval and Early Modern Medicine
ISBN 978-3-031-69252-9 ISBN 978-3-031-69253-6 (eBook)
https://doi.org/10.1007/978-3-031-69253-6

Cover credit: Giorgio Morara

This Palgrave Macmillan imprint is published by the registered company Springer Nature Switzerland AG
The registered company address is: Gewerbestrasse 11, 6330 Cham, Switzerland

Acknowledgments

As a concerted effort, this edited book would not have been possible without the contributions and support of many individuals. Dante once said, "Diverse voci fanno dolci note" [Differing voices join to sound sweet music, *Pd.* 6.124], and I truly hope that the "dolce armonia" of this endeavor resonates through the pages of this book. My deepest gratitude goes to Fabrizio Bigotti (Julius Maximilian University of Würzburg) and Jonathan Barry (University of Exeter) at the Centre for the Study of Medicine and the Body in the Renaissance—Fondazione Comel (Pisa, Italy), for inviting me to develop this project as part of their commemoration of the seven-hundredth anniversary of Dante's death in 2021. Despite the constraints of international COVID-19 protocols, they successfully coordinated an exciting online conference, which evolved into the present volume. I am grateful for their generosity in hosting this work in their book series.

I extend my sincere thanks to Andrea Celli (University of Connecticut), Ivano Dal Prete (Yale University), Eva Del Soldato (University of Pennsylvania), Christiana Purdy-Moudarres (Yale University), and Arielle Saiber (Johns Hopkins University) for accepting my invitation and contributing their valuable research. I am also grateful to Tomaso Maria Pedrotti Dell'Acqua for facilitating the organization of the event, and to all the attendees who enriched the discussion on Dante and medical thought.

I would like to acknowledge my profound indebtedness to the scholars who have guided my research on medieval medicine and literature, both within and beyond my doctoral program. Teodolinda Barolini (Columbia University), Joel Kaye (Barnard College), and Pier Mattia Tommasino (Columbia University) have been insightful readers of my work, offering their time and expertise with unmatched generosity.

The title of this volume is inspired by a panel I organized at the 53rd International Congress of Medieval Studies at Kalamazoo (Western Michigan University, May 10–13, 2018), sponsored by the association Italians and Italianists at Kalamazoo. I am grateful to Akash Kumar (University of Berkeley) for including it in the conference program. This conversation continued at the 57th International Congress of Medieval Studies at Kalamazoo (Western Michigan University, May 9–14, 2022), and at the virtual Renaissance Society of America meeting (November 30–December 3), with two panels sponsored by the Dante Society of America. My thanks go to Dr. Kumar once again, the DSA liaison at Kalamazoo, and Filippo Gianferrari (University of California, Santa Cruz), the DSA liaison at the RSA, for accepting the panels, as well as to all the presenters who shared their work. Special thanks to Catherine Bloomer (Brandeis University) and Joseph Romano (Columbia University), who eventually contributed chapters to this volume.

Lastly, I am deeply thankful to the editors at Springer Nature-Palgrave Macmillan (Lynnie Sharon, Lucy Kidwell, Eliana Rangel, Birke Dalia, Petra Treiber, and Seethalakshmi Vijayasarathy) for their meticulous coordination of the production of this volume, and to the anonymous reviewers who provided insightful feedback on the proposal. As always, any remaining errors rest solely on my shoulders.

Praise for *Dante and the Sciences of the Human*

"The essays collected in this volume enrich our knowledge of Dante's understanding and poetic use of natural philosophy by exploring from different perspectives a crucial topic in medieval intellectual history: the medical, theological, and political perception of the human body as a new scientific consciousness emerged. In exploring Dante's corporal imagery by placing in dialogue different disciplines, this book offers an original and stimulating contribution to our understanding of the unity of all fields of medieval knowledge."

—Susanna Barsella, *Fordham University*

"The focus of this fine collection of articles on Dante's 'sciences of the human,' built through his seamless integration of field after field of contemporary knowledge—natural philosophy, medical theory, and theology, among others—reveals anew the startling interdisciplinary breadth of his intellectual project."

—Joel Kaye, *Barnard College*

"*Dante and the Sciences of the Human* constitutes a timely intervention that contributes to broader intellectual discourse extending beyond the field of Dante Studies. By delving into the nuances at the crossroads of poetry and natural sciences, this book not only presents a series of innovative interpretations of Dante's works but also imparts valuable insights into their enduring significance across disciplines. It also emphasises that no reading occurs in isolation, thereby presenting a compelling argument for Dante scholarship to integrate multiple areas of knowledge, particularly the sciences. In doing so, *Dante and the Sciences of the Human* leads to a deeper understanding of the complexity of Dante's works and their significance within the extensive history of scientific thought in relation to concurrent developments in the humanities."

—Eugenio Refini, *New York University*

Contents

Notes on Contributors

Catherine Bloomer is Visiting Assistant Teaching Professor of Italian Studies at Villanova University. Prior to her current appointment, she was the Florence Levy Kay Fellow in Premodern Disability Studies at Brandeis University. Bloomer is a scholar of medieval Italian literature and premodern disability studies in the Mediterranean world with a specialization in gender and the body. She is Associate Director for The New School art education program WriteOn NYC and Assistant Editor for Digital Dante, an online repository for original research and ideas on Dante and his work. Bloomer's work combines textual analysis, material culture, and disability theory to examine the history, representation, and treatment of disability in premodern texts.

Andrea Celli is Associate Professor of Italian Literature and Cultural Studies at the University of Connecticut, where he teaches Italian and Mediterranean Studies. In addition to several books, articles, and translations from Arabic and French authors, he recently published *Dante and the Mediterranean Comedy: From Muslim Spain to Postcolonial Italy* (Palgrave Macmillan, 2022). He is currently working on two book projects: a monograph on early-modern representations of Hagar and Ishmael, as genealogical symbols of religious conflict; and a book on Muslims and the Eucharist, from medieval legends to modern medievalism.

Ivano Dal Prete is Senior Lecturer of History at Yale University. His research focuses on the history of early modern Earth sciences, astronomy, popular science, and scientific communication networks. He is the author of *Scienza e Società nel Settecento Veneto* (FrancoAngeli, 2008). His most recent book, *On the Edge of Eternity: The Antiquity of the Earth in Medieval and Early Modern Europe* (Oxford University Press, 2022), questions the common narrative of an eighteenth-century discovery of geological time, arguing instead that the idea of an ancient Earth was widespread and freely discussed in medieval and early modern Europe.

Eva Del Soldato is Associate Professor of Italian Studies at the University of Pennsylvania. Her research is primarily devoted to Renaissance thought and culture, with a special attention to the Aristotelian and Platonic traditions. Alongside several articles and editions, she is the author of the monographs *Simone Porzio* (Edizioni di Storia e Letteratura, 2010) and *Early Modern Aristotle: On the Making and Unmaking of Authority* (University of Pennsylvania Press, 2020). She was also the recipient of fellowships from the Scuola Normale Superiore di Pisa, Villa I Tatti, the Herzog August Bibliothek in Wolfenbuettel, and the Huntington Library in Pasadena.

Matteo Pace is Assistant Professor of Italian Studies at Connecticut College. Upon earning a PhD in Italian and Comparative Literature from Columbia University, he won a Santorio Award for Excellence in Research from the Centre for the Study of Medicine and the Body in the Renaissance, where he is also an Associate Fellow. In his research, he focuses on the intersections between vernacular cultures and medical thought in the thirteenth and fourteenth centuries. Besides Dante, he has published on Boccaccio's *Decameron*, Guido Cavalcanti and Dino del Garbo (*Studi sul Boccaccio*, 2016; *Mediaevalia*, 2025), on Giacomo da Lentini and the Aristotelian tradition (*Traditio*, 2020), on Guido Guinizzelli, Avicenna, and Taddeo Alderotti's medical philosophy (*Romanic Review*, 2022), and on Catherine of Siena's theology of blood (*Italica*, 2022). His current book project explores the varied networks of medieval literary cultures through the cultural reception of ancient and medieval medicine.

Joseph Romano is a PhD candidate in the Department of English and Comparative Literature at Columbia University. His research focuses on the reception of scholasticism in Dante and Middle English literature, regarding especially questions of love, soul, and volition. His dissertation,

"Stirrings" of the Will: Issues of Volition in Late 14th-Century English Literature, develops a history of the will in Middle English poetry and works of contemplation.

Arielle Saiber is Charles S. Singleton Professor of Italian Studies at Johns Hopkins University. Saiber publishes primarily on medieval and early modern Italian literature, the history of mathematics, and science fiction. Her books include *Giordano Bruno and the Geometry of Language* (2005), *Measured Words: Computation and Writing in Renaissance Italy* (2017), and the co-edited anthology *Images of Quattrocento Florence: Writings on Literature, History and Art* (2000). She is co-editor of the web archive *Dante Today: Citings and Sightings of Dante in Contemporary Culture*, and of the University of Minnesota's new interdisciplinary monograph series, *Proximities: Experiments in Nearness.*

Abbreviations

Civ. Dei	*De civitate Dei*
Conf.	*Confessiones*
Conv.	*Convivio*
De gen.	*De generatione et corruptione*
DVE	*De vulgari eloquentia*
Ep.	*Epistola*
Inf.	*Inferno*
Met.	*Metaphysica*
Mon.	*Monarchia*
Par.	*Paradiso*
Purg.	*Purgatorio*
Rvf.	*Rerum vulgarium fragmenta*
SCG	*Summa contra gentiles*
ST	*Summa theologiae*
Tr. Cup.	*Triumphus cupidinis*
Tusc. Disp.	*Tuscolanae disputationes*
VN	*Vita nuova*

List of Figures

CHAPTER 1

Dante's Sciences of the Human: An Introduction

Matteo Pace

Despite the otherworldly qualities of Dante's journey, the *Commedia* is brimming with bodies—"sì lunga tratta / di gente," that the poet is at pains in believing "che morte tanta n'avesse disfatta" (*Inf*. 3.55–57).[1] This may seem a prosaic consideration, but without those bodies—their punishments, their earthly lives, their tears, laughs, their human passions and ailments—the *Commedia* would present a much different landscape. Across the poem, Dante encounters the incorporeal bodies of the damned and the blessed with all their *corporeal* consistency. Despite being unable to embrace the *ombra vana* of Casella in *Purgatorio* 2, he can nonetheless walk over the "vanità che par persona" (*Inf*. 6.36) of the gluttonous.[2] Such inconsistencies in the physicality of the souls are commonplace in Dante's studies. Yet, they do not deter Dante from investigating the bodily realities of his textual creation with scientific precision unmatched in medieval Italian literature.[3] By taking the title

M. Pace (✉)
Department of Italian Studies, Connecticut College, New London, CT, USA
e-mail: mpace1@conncoll.edu

M. Pace (ed.), *Dante and the Sciences of the Human*, Palgrave Studies in Medieval and Early Modern Medicine,
https://doi.org/10.1007/978-3-031-69253-6_1

of this collective volume at the core of my research questions, in this introductory chapter, I explore how Dante constructs a cogent relationship between the sciences of the human in his work. How does the issue of corporeality—understood as the theoretical and material springboard for an intellectual discussion encompassing medicine, natural philosophy, and religious discourse—imbue Dante's poetic and philosophical achievements? How does the body, considered both as a category of scientific investigation and as a narrative construction, represent a line of inquiry in the interpretation of Dante's work across the centuries? In sum, how and why did Dante look at bodies in his work, and how did bodies look back at Dante through the intricate web of his reception?

Purgatorio 3 stands out in the *Commedia* as a representative instance of how Dante's reflection on the body permeates his thought. On the shores of the Ante-Purgatory, after Catone scolds the souls for lingering on Casella's beautiful singing, Dante resumes his journey with a troubled Virgilio at his side. The flaming red rays of the sun on his back make his body cast a shadow on the ground, a physical experience that underscores his corporeal presence.[4] Dante sees only one shadow in front of him ("solo dinanzi a me," *Purg*. 3.21) and is alarmed ("con paura," line 19) at the mere thought that Virgilio has abandoned him—an eerie foreshadowing of Virgilio's departure later in the canticle, just as Dante is embarking on the final leg of his journey. His human reaction, depicted with such a natural touch, carries a philosophical gravitas that holds the entire canto together. Since Dante cannot experience Virgilio's body next to his, his body is understood as the objective correlative of his sensible doubting. The immateriality of Virgilio's shade—unable, as it were, to cast a shadow next to Dante's—forces readers to question knowledge without physical sensation. The core of *Purgatorio* 3 challenges thus human sensitivity and the certainty of presence without physical consistency, foreshadowing even more rarified representations in the heavens of *Paradiso*.[5]

Virgilio's response to Dante, reassuring him of his presence, transcends the physicality of their human exchange. It is truly *metaphysical*—μετὰ, "after" or "beyond," and τὰ φυσικά, "all things natural"—yet so bodily in the way his companion still understands it:

> e 'l mio conforto: «Perché pur diffidi?»,
> a dir mi cominciò tutto rivolto;
> «non credi tu me teco e ch'io ti guidi?

Vespero è già colà dov' è sepolto
lo corpo dentro al quale io facea ombra. (*Purg*. 3.22–26)[6]

Dante's trust in Virgilio should go beyond the physical reassurance of sensible apprehension, at this point in the journey. Virgilio's response—and trust—is limited by his being *al di qua* of Christian revelation, as his following explanation makes evident. His own understanding cannot fully grasp the reality of his shade: "A sofferir tormenti, caldi e geli / simili corpi la Virtù dispone / che, come fa, non vuol ch'a noi si sveli" (lines 31–33).[7] This conundrum, awaiting further explanation by the Latin poet Stazio—a closeted Christian, in Dante's fiction—deepens the abyss keeping Virgilio on the other shore of salvation, under the physical mark of the body. In merging philosophical and literary modes, Virgilio's realization of his own intellectual (and bodily) finitude mirrors Bonagiunta Orbicciani's towards the end of the purgatorial journey, the thirteenth-century poet marveling at Dante's renewed style: "colui che fore / trasse le nove rime" (*Purg*. 24.49–50), acknowledging the surpass of the Stilnovo generation, and "il nodo / che'l Notaro e Guittone e [lui] ritenne / di qua dal dolce stil novo" that he hears (lines 55–57).[8]

John Freccero (1986: 195–208) noticed intratextual connections between poetic and biological creation, focusing on bodily features in *Purgatorio* 3, read against the philosophical and literary history developed across *Purgatorio* 24 and 26. As Dante and Virgilio climb their way up on the mountain, they encounter the soul of Manfredi, the natural son of Frederick II and last hope of the Ghibelline forces in Italy. The group of shades that they meet marvels at the sight of Dante's shadow, opposite to his earlier fear of Virgilio's hastened departure. Although "biondo era e bello," Manfredi's face is disfigured by "un colpo" that left a scar on his brow (*Purg*. 3.107–08).[9] Freccero stresses the "anomaly" of such a representation (1986: 196), as a physical sign of the soul's bodily death still carrying sensitive weight in the afterlife. Over and over again, Dante indulges in a reflection on mimesis that brings textual reality in contact with philosophical representation.[10] Manfred's wound, with all his fictional and theological anomaly, allows Dante for a series of parallels that draw him closer to the Gospel of John, where Christ's wounds in the hands of Thomas testify to the very reality of his bodily resurrection. John's text "demand[s] of the reader the same assent that is demanded of the doubting Thomas" (Freccero 1986: 200), and by the same token, Dante's fiction forces readers to question the physical reality of his shades beyond the boundaries of sensible perception. *Purgatorio* 3, its shadow cast onto the scientific precision of canto 25 and the poetic manifestos of cantos 24 and 26, makes the *Commedia* a text where bodies are central to literary and scientific representation. At once surpassing and creating a literary history of thirteenth-century vernacular literature, Dante signals how philosophical and textual barriers merge a new dimension of the body into a novel science of the human.

Medicine in Dante's Biography

Dante's dimension of the bodies—his fictional creations, the medical and philosophical reflections on their reality, the theological underpinnings—constitutes the focus of this volume. As the subtitle of this book conveys, the *sciences of the human* encompass medical and scientific thought, natural philosophy, and theological writings, in order to show the interdisciplinary breadth of Dante's own intellectual project. Dante's engagement with encyclopedic knowledge allowed him to further connections between disparate spheres of investigation, which nurtured his poetry and his philosophical language across his oeuvre. Critical assessments of Dante's exposure to medical and scientific thought underline the depth and solidity of his approaches, standing out in the literary panorama of thirteenth- and fourteenth-century Italian literature as "not at all predictable" (Tonelli 2021: 227). Dante's expertise is evident from his first literary masterpiece, the *Vita nuova*, where human physiology and pneumatology take the front stage in describing stages of lovesickness. As Tonelli argues, Dante's self-fashioning as a "visionary poet-figure," with evident "prophetic" qualities—which will have so much importance in the authorial reality of the *Commedia* (Nardi 1949: 336–416; Barolini 1992: 3–20)—is already well defined in the *libello*, part of an "intellectual self-portrait" that takes its lymph in "a certain 'humoral determinism'" (Tonelli 2021: 227). Dante's relationship with medical thought and natural philosophy is thus one of the major threads of his intellectual formation, and a relevant step in the development of medicine in medieval Italy. Already in the *Vita nuova*, Dante is in dialogue with theories of embodiment not as mere erudite references to embellish his poetic production, but as constitutive parts without which the overall sense of the *libello* would be incomprehensible. By fashioning himself as a visionary poet via a scientific understanding of humoral determinism, Dante centers on the physicality of his poetry—from the material cognitive functions that open the book of memory, to the physiological manifestations of love, and the prophetic qualities of his connection with Beatrice.

Whether or not Dante directly examined the medical treatises of Hippocrates, Galen, Avicenna, and other authoritative writers discussed in the European *Studia* of medicine, their contents were part of the encyclopedic writings in circulation at the time, *summae* that offered the chance to connect specialized points in human physiology with wider arguments about the relationships between micro and macrocosm.[11] In this perspective, Dante's original contribution to questions of embodiment must be evaluated alongside medieval cultures, continuously engaging with the sciences of human and literature. As scholars devoted to the intricacies of Dante's biography have already shown, among the many lacunae that constituted his documented life, one fact seems to remain unquestionable. Sometime in the latter half of 1295, Dante enrolled in one of the major guilds of the Florentine *comune*, the Guild of Physicians and Apothecaries—a necessary step toward a political career in Florence in the tumultuous years that led to his eventual exile from the city (Inglese 2015: 15, 61–68).[12] Although his professional commitment to the more practical business of healthcare must have been limited—if not downright null—Dante's relationship with the scientific environment of his time is evidence not only of the philosophical core of contemporary medical learning but also of the profound connections between university culture and literary movements during the Stilnovo years.[13]

During Dante's formative years, a culture coalesced around Bologna's intellectual environment and its university, not only on account of philosophizing trends within medical thought, but also on account of its engagement with interdisciplinarity and vernacular literature. Interdisciplinary as the interests of Taddeo Alderotti and his pupils, who commented not only on medical treatises, but also moral, economic, and literary texts (Siraisi 1981; Chandelier 2017; Duranti 2022). Between Bologna and Florence, a vernacular intellectual class emerged, comprising academics and physicians like Dino del Garbo, who wrote the first medical commentary on Guido Cavalcanti's doctrinal canzone on the nature of love, *Donna me prega* (Fenzi 1999), or men of the law like Guido Guinizzelli, who, in the same years when Taddeo Alderotti was revolutionizing the teaching of medicine in Bologna, discussed lovesickness and social capital in his poems (Borsa 2007; Pace 2022). Jacobus de Pistoia, magister in the same *Studium*, dedicated his *Questio disputata de felicitate* to Dante's *primo amico* Guido Cavalcanti, exemplifying and solidifying the web of relations between cities, university culture, vernacular poetry, medicine, philosophy, and ethics (Kristeller 1955; Corti 1982; Corti 1983; Zavattero 2005). A university city, Bologna, that was not foreign to Dante himself, who stayed there sometime before 1287, when one of his early sonnets was transcribed in the *Memoriali bolognesi* (Orlando 1981; Antonelli and Pedrini 2001; Bellomo 2004; Antonelli 2010; Antonelli 2018)—in a time, then, when Florence entertained many professional and cultural relationship with the city, especially in the political entourage of Dante's own family (Inglese 2015: 37–38). Although documentary evidence is somewhat lacking, Dante might have been in Bologna again in the early 1290s, and possibly during the first years of his exile, when he was attending to the philosophical treatises of the *Convivio*, in the years 1303/04–1307/08 (Fioravanti 2014: 5–79), and the *De vulgari eloquentia*, in the years 1304–1306 (Tavoni 2011: 1067–1116).[14] Dante's connections with university cultures thus connect Florence with Bologna, their literatures and sciences engaging in cogent conversations around human materials and intellective components, with their poetic representations and hermeneutics. A community of intents that nurtured Dante's poetic development from his first literary experiments, which take part in his personal shaping of embodiment.

Medical Readings Through Dante's Readers: A Critical Appraisal

Medical and scientific readings of Dante are not a critical novelty, nor a prerogative of Dante studies. As mentioned earlier, in the same decades as Dante was composing his philosophical treatises and the *Commedia*, the physician Dino del Garbo commented upon Guido Cavalcanti's canzone *Donna me prega*, a doctrinal and challenging poem that investigates the nature of love from a philosophical standpoint. Dino's commentary reads through the canzone as if it were a medical treatise to be taught in school, by dividing the text as a Scholastic *lectio*, and by recognizing Cavalcanti's *altezza d'ingegno* in matters of physiology and natural philosophy. As it were, Dino states that Cavalcanti spoke "scientifico modo et veridico, tracto ex preceptis scientie naturalis et moralis" (Fenzi 1999: 88), thus fully crediting the scientific and philosophical potential that is inherent to poetry. Similarly, scientific credit was given to Dante's *Commedia* almost at the same of its publication, if one considers the amount of critical efforts that the first commentary tradition exerted in deciphering and explaining the most intricate aspects of Dante's philosophical, scientific, and theological background. In matters closer to medical discourse, the earliest commentaries already provide notes on the authorities mentioned or met by Dante in the course of his journey through the underworld (Prioreschi 1994). Carmelite friar Guido da Pisa, among the first to comment on Dante's *Inferno*, mentions Galen as the author of the *Liber de interioribus* and many other treatises ("plus quam C volumina"), and refers to him as "medicus," with the same antonomasia that pertains to Aristotle "philosophum," or Vergil "poetam" in *Inferno* 4.[15] As an erudite and passionate reader of the poem, Guido is willing to uncover the rhetorical and historical precisions of Dante's references in the *Commedia*, and to give his scientific expertise the credit it deserves.

In the third quarter of the fourteenth century, scholar Benvenuto da Imola writes his own commentary to Dante's *Commedia*, and introduces erudite references into the exegesis of the poem, once again emphasizing medical and scientific matters as an integral part of Dante's poetic discourse. In glossing the precision with which Dante weighs "il sangue ferrarese" lamented in *Par.* 9.55–57 as a tiresome and wretched task, measuring blood "a oncia a oncia" (line 57), Benvenuto recognizes Dante's medical competence ("et verum dicit") through Galen's anatomical authority ("unde Galenus scribit"). By illuminating the poem through its scientific references, Benvenuto uses medicine as the sole background capable of fully uncovering the tragedy expressed in the tercet.[16] Furthermore, in the final vision of *Paradiso* 33, Benvenuto uncovers the medical exactitude with which Dante compares the rapidity of successive images in the face of god, with his mental incapacity to hold them in memory, as a "letargo" (*Par. 33*.94) longer than the chronological distance of twenty-five centuries. In extending the medical reference beyond the scientific background of other commentators, Benvenuto thus shows how Dante's medical references are organically embedded into the rhetorical power of his poem and the intellectual breadth of his literature.[17]

Benvenuto's and Guido's methods of reading Dante's scientific context as part of the meaning of the whole text are not scattered instances.[18] Dante's own son Pietro, who commented on the poem in more than one instance during his life, suggests that his readers will be able to understand the allegorical blindness of *Paradiso* 26 only through a careful understanding of the anatomy of the eye, taken verbatim from Galen's anatomo-physiology. By glossing the simile of *Par.* 26.70–75, and in particular the technical word "gonna" (line 72), Pietro evidences Dante's contribution to the scientific vocabulary of the vernacular, whence *gonna* translates the Latin "tunica cornea," and connects his father's medical trope to the intellectual vision that he is trying to convey in his poem.[19] Through the earliest exegetical efforts that unveiled the *Commedia*'s medical vocabulary and precision, fourteenth-century readers of the poem were already able to understand not only the solidity of Dante's scientific culture but also the interdisciplinary connections that constitute his science of the human.[20]

As this collective volume grapples with a multifaceted approach, integrating various branches in the field of medieval studies, a summary (and thus, incomplete) of significant scholarly contributions in modern times serves as an introduction to the main questions addressed. Connections between medical thought and human sciences constitute a "never forgotten, nor muted, consanguinity… between medicine and the humanities," as a recent anthology of Dantean scholarship has highlighted (Di Nuzzo 2021: 8).[21] During the nineteenth and early twentieth centuries, erudite physicians (especially those working on the historical side of their discipline) read Dante's *Commedia* "from a diagnostic viewpoint" (8), in order to emphasize both the significance of Dante's borrowings from contemporary scientific cultures, but also to address the deep significance of the physical and the material in his poetry. From the physiognomic of Cesare Lombroso, who investigates the epileptic nature of Dante's neurosis, to the psychiatric analyses of Bernardo Chiara in response to his friend Lombroso, the interpretation of the poem cannot eschew a sustained confrontation with both the thorniest aspects of his biography, and the contextualization of medical philosophy during his life.[22] Historical reconstruction thus offered a fertile field of investigation to bring significance to the most complex aspects of Dante's erudition, as posited at the beginning of the twentieth century by historians and practitioners of medicine like Arturo Castiglioni, Liborio Giuffré, Antonino Del Gaudio, and Libertino Alajmo.[23]

As we modern readers strive to unveil the intellectual history of Dante's times, represented by the poet in his literary and philosophical oeuvre, we need to recognize the *extravagant*—i.e., diverging from the main threads of its tradition, as Gianfranco Contini would put it—approaches that have been part of Dante's criticism since its very establishment. And in reckoning with this history, such a critical attitude would not have developed in recent years without the philosophical acumen of historian Bruno Nardi, who dedicated a great deal of his scholarship to reading Dante and medieval literary cultures as part of larger intellectual environment. Since his Leuven dissertation on Siger of Brabant and Dante's philosophical sources (1911, 1912), and through the rediscovery of Albert the Great's contribution to medieval thought as one of Dante's favored authorities (1922), Nardi's approach to Dante's work has contributed to put his literature on the philosophical map of the Middle Ages.[24] Among his vast bibliography on Dante studies, Nardi unveiled some essential scientific and medical debates in the *Commedia*, and the rest of his lyrical and philosophical production.[25] Through an exact close reading of literary texts, Nardi assessed as well the scientific awareness of many authors of Dante's generation, and enriched the conversation in terms of precise hermeneutical problems. By unveiling the lexical richness of medieval Italian poetry, in and outside of Dante, Nardi's historical research in the field of philosophy and science was able to define how the sciences of the human consciously put literature, medicine, and university cultures in communication with one another.

Following the steps of Nardi, scholars have continued to unearth the richness of Dante's intellectual preparation in numerous groundbreaking directions. Sonia Gentili's work (2005) on Aristotelian ethics and anthropology in thirteenth- and fourteenth-century Italian literature managed to focus the conversation on numerous authors and works—Dante, of course, but also Guido Cavalcanti, Francesco Petrarca, the collection of short stories known as the *Novellino*, and moral literature—and their contextual cultural framework. As Gentili argues, the philosophical debates of these centuries cannot eschew the scientific engagement of the masters of arts, and especially the physicians, who acted as true innovators—and often as mediators—of philosophical culture to and from the vernacular. This approach was particularly productive in her studies on Taddeo Alderotti and his vernacular translation of the *Summa alexandrinorum*, where she addresses issues in which medical thought is in communication with a whole cultural panorama, including vernacular debates in poetry and prose that were particularly close to Dante's heart (2004, 2006, 2012a, 2012b, 2014).

Critical inquiries on Dante's philosophical background have shown how the investigation on sources—whether or not accessed through a direct examination of the texts, or through the widespread *summae* of the time, collections of authoritative excerpts largely used in the Middle Ages—must always be connected with a hermeneutic drive that goes back to the letter of the text. As in the studies of Maria Corti (1982,

1983) and Maria Luisa Ardizzone (2011, 2016), the historical assessment of philosophical and scientific treatises commented in the schools brings forth a host of interpretive problems on the significance of Dante's word, often challenging received notions, and stimulating the critical debate. Medical thought as practiced in the university *Studia*, always in conversation with the philosophical trends of Aristotelianism, Galenism, and the Greek, Hebrew, and Arabic traditions of scientific commentaries, nurtured the critical efforts of scholars who investigated the connections between Dante and his intellectual panorama. Medicine is thus in constant conversation with both theology and natural philosophy, bringing the lexical technicalities of Dante's oeuvre at the forefront of this comparative perspective.[26] In this vein, Vittorio Bartoli's and Paola Ureni's focus on Galenic medicine unpacks Dante's debts towards medical thought by interrogating psycho-physiological questions through their relation on human faculties, and thus indicating—as does Tonelli with respect to Dante's prophetic instances (2015, 2021)—how intertwined the sciences of the human were to the disparate corners of the medieval cultural landscape.[27]

This comparative methodological approach requires the concerted effort of scholars coming from different fields of studies, as a recent collection of essays edited by Gaia Gubbini (2020) made all the more evident. While inviting a conversation between literary critics, philologists, historians of medicine, and of science, the volume reinforces the idea that "medieval cultures had much more permeable borders" (2) than siloed academic fields can contain. A psychosomatic inquiry into the Middle Ages thus presents lines of investigation that cannot be easily compartmentalized, and brings forth an existential conundrum for the immediate future of medieval studies, where disciplinary specialties require a continuous communication among themselves.

As readers were keen to understand early on in the reception history of the *Commedia*, the substance and depth of Dante's engagement with medical and scientific literature constitute a privileged thread in the interpretation of his work. Building upon the identification and deconstruction of this source material, Dante's readers were faced not only with the task of unveiling the exactitude of his scientific mind, but they had to engage themselves with the whole significance of Dante's sciences of the human. Bridging historical research with literary criticism opened the gates to reading Dante's polysemous oeuvre from multiple perspectives.

This perspective can thus offer some methodological points to illuminate the approaches followed in this whole volume. In a very recent book by Giorgio van Straten (2023), the relationships between historical documents and literary work are questioned through the lens of the historian-turned-literary-critic, and vice versa. While the statement that literature certainly has the status of a historical source sounds like a trivialization of a deep engagement and analysis, the methodological approach to the use of literature in historiographical research is a question that involves the literary historian as much as the *historian* historian. Van Straten underscores that the mere dichotomy between truth and falsehood, which determines what is accurate and what is fictional in literature, does not wholly encompass the ways in which literary sources can contribute fruitfully to historiography. Literature's value lies not only in its content, but also in its specific language—the linguistic choices of authors, their depiction of reality, and even what they choose to omit. Thus, the form of literary works—whether philosophical prose or poetic expression—shapes their argumentation. Literature can transcend the limitations to which it is seldom confined, allowing historians not only to use it as a source, but also to engage with its unique epistemology and its elaboration of reality. Through this lens, historians and literary critics can accost their analyses through the pages of their authors and documents alike.

As the following chapters grapple with these questions from their own, privileged points of view, we might wonder how this insight can be revealing of Dante's deep engagement with medicine and human sciences. Although van Straten addresses mostly modern and contemporary literature and history, his arguments have a certain degree of resonance for many scholars involved with literature and history, and the chapters in this book show how Dante's poetry and philosophy are a fundamental step in understanding the intellectual history of thirteenth- and fourteenth-century medicine and science. The authors contributing to this collection posit that the relationship between Dante and medieval cultural history is not exhausted by the investigation of the sources, nor by the transmission and/or correct interpretation of a single treatise. Dante's sciences of the human—medicine, science, theology—reveal his approximation of a human theory, a vision that nurtures his poetry and philosophical literature, but also his permanence as an authority in human sciences through the centuries. Using the categories proposed by Maria Malatesta (2005) in her reading of Hans Fallada, Dante is thus *witness* and *representation* of a multifaceted reality, where his sources—medical theories, natural philosophical arguments, religious underpinnings—and their (re)presentation are in constant dialogue, as is Dante's historical reality with his own perception and categorizations.

An Overview of This Volume

One of the focal points of this collection foregrounds the theoretical relationship between literature—in the case of Dante, poetry and philosophical treatises—and history, particularly the intellectual history of thirteenth- and fourteenth-century Italy. *Dante and the Sciences of the Human* proposes a comparative reading of Dante and his times, by approaching his contribution to medical, scientific, and spiritual thought in medieval and early modern times, through an interconnected lens. As the following chapters demonstrate, in his poetry and philosophical writings Dante shaped an understanding of the human body and mind through a deep engagement with medical and scientific thought. Dante's multifaceted approaches to the physical account for an interdisciplinary standpoint that puts medieval literature in conversation with the history of medicine and the body, politics, theology, philosophy, and Dante's reception through the centuries. The authors of the chapters are continuously responding to one another, with a concerted effort that shows how Dante's questions on science, nature, and the soul are actively in conversation across different times, texts, and geographical spaces.

The richness of these conversations can be explored through various axes, without curtailing further potential conjunctures that may arise while reading these pages. These introductory pages (this chapter) frame Dante's interdisciplinary breadth between literature, medicine, and theology, while also proposing how a science of the human constitute a privileged thread in investigating Dante's work, and a recognizable interpretive trend since the earliest commentary tradition of the *Commedia*. The volume continues this discussion by focusing on Dante's relationship with human placement, understood either as the place of human life in the macrocosm, or the place of human emotions in the body/soul debate. Ivano Dal Prete's "When the Ocean Towered Above the Earth: Dante's *Questio* in Context" (Chapter 2) shows how Dante's *Questio de aqua et terra* responds to a wider historical and scientific debate that questions the place of the earth in the created universe, while in "Passion and Passionless Love in Dante: from *Epistola* 3 to the *Commedia*"

(Chapter 3), Joseph Romano posits that Dante's physical placement of human passions pushes physiology and philosophical authorities toward unprecedented directions. Dal Prete contextualizes Dante's *Questio* in the scientific and geographical knowledge of his time, by positing that Dante's treatment of the apparently illogical statements on the relative positions of earth and water furthers a deep engagement with scientific reasoning, even when it contrasts with the immediate perception of the senses. The *vexata questio* of the contradictions between the treatise and the *Commedia* are thus two faces of the same medal, which are by no means mutually exclusive in Dante's mind, nor in his intellectual context. Poetry and science therefore are equally able to provide solid explanations through their own language, be it meteorology or poetic fiction. By the same token, Romano's investigation on volition and passion between the *Commedia* and the rest of his production shows how Dante varying positions account for a slowly maturing realization of the relationships between human and divine will. In this perspective, human materiality is thus a battlefield for poetic, theological, and philosophical debates on the placement of the humans in the created universe. By exemplifying Dante's encyclopedic ability to put into communication disparate spheres of knowledge, Dal Prete and Romano show how a reflection on Dante's sciences of the human brings both his poetry and treatises to the forefront of philosophical sophistication.

Dante's relationship with social and political issues focuses on an embodied perspective that takes into account medical thought and human sciences, which are the foci of the following two chapters. In "Anatomy of Florence: Illness and Body Politic in the *Commedia*" (Chapter 4), Andrea Celli offers a comparative reading of Dante and al-Fārābī to show how the debates on self-governance of the city and the state push medical metaphors into the sphere of social tensions, while in her "Neither O nor I: The Handiwork of *Inferno* 24 and 25" (Chapter 5), Arielle Saiber argues that Dante's punishment of the thieves questions their ontological status as created beings. Through a contextualization of Dante's political theory from an embodied perspective, Celli proposes that the *Commedia* offers an understanding of the city in human physiological terms, and is therefore in communication with analogous allegories of body politics derived from classical philosophy and its Arabic and scholastic interpreters. Between the poem and his philosophical production, Dante thus engages with Florence from a scientific and moral standpoint, which allows him to identify the causes of its political failures via a medical perspective. The ever-changing political landscape of the *comune* is further illuminated by a comparative reading of Abū Naṣr al-Fārābī's treatise on the virtuous city, which shows how Dante's perspective on these crucial questions cannot eschew the organic imagery therein, and the scientific backbone of medieval political theory. While engaging with scriptorial practices of thirteenth-century manuscripts, Saiber offers a reading of *Inferno* 24 and 25 that contextualizes the *contrapasso* of the thieves within the images of the cantos. Epistemological and ontological reflections between object and thing nurture her navigation through the *terzine* of the cantos, and show the stakes of a physiological understanding of human identity. Writing is then another embodied feature of human sciences, developed by Dante as poetic imagery that accounts for the hellish condemnation of the thieves. The ambiguity of Dante's words opens up a paleographical speculation on letters and numbers, which frames the shifting bodies of the damned deep within the folios of the manuscript. Physiology and poetic are then put in proximity thanks to the development of medieval algebra, where the metamorphoses of the thieves zero onto their wavering sense of the self. In putting Dante's *Inferno* in conversation with body politics and embodied writing, Celli and Saiber illuminate Dante's understanding of identity within the fibers of the community, and show the extent to which a disruption of balance can affect human relations.

While questions of reception have been addressed in several of the previous pages already, the final two chapters investigate how readers of Dante use him as a scientific and philosophical authority in two different Renaissance contexts. Through a reconstruction of Dante's (often oblique) presence in sixteenth-century Florence, Eva Del Soldato's "Speaking Like a Philosopher: Dante and Renaissance Love Treatises" (Chapter 7) reveals how Dante's thought, and his relationship with medical philosophy, reverberate through Renaissance debates on human love, while Catherine Bloomer's "Disabled Body and Female Soul: from Dante to Marinella" (Chapter 6) reads Lucrezia Marinella's abundance of Dante's citations in *La nobiltà e l'eccellenza delle donne*, to show how a Renaissance approach to human bodies complicates the intellectual understanding of soul and matter from a gendered perspective. Del Soldato's study on the Florentine *accademici* engages with Dante's authority as both philosopher and poet, by positing that these features cannot be easily separated. Through an analysis of many lectures on the *Commedia* and his lyric production, given by the members of the

Accademia Fiorentina, Dante's oeuvre is often in direct communication with Petrarca's physiology of passion, and casts a long shadow on a medical and organic understanding of human love. On a different trajectory, Lucrezia Marinella's treatise on female excellence painstakingly rewrites texts and contexts of Dante's *Commedia*, in order to push forward a science of women's souls and bodies that has to be put in conversation with his *Convivio* just as well. In gesturing towards Dante's statements on intellectual and physical disabilities, Marinella's positions on women's bodies turn the misogynistic claims of Giuseppe Passi onto his head, and reclaim a more accurate reading of Dante's philosophy of the body in lights of the Renaissance *querelles des femmes*. From both their perspectives, Del Soldato and Bloomer argue that Dante's understanding of the body, his functions and ethical goals have pivotal consequences in the ways these topics are discussed in the later centuries, particularly when engaging with medical thought.

The stimulating insights this volume aims to offer are just one example of Dante's rich understanding of the interconnected realities of bodies in medieval culture. By bridging the fields of medicine, literature, and theology, Dante's works stand out in the Middle Ages as offering a truly multidisciplinary perspective on human embodied identities. His poetic and philosophical imagination is ever engaged with the presence of bodies, which also constitute an integral part of his scientific understanding of the created world. By exploring the corporal imagery in the *Commedia*, as well as his political and encyclopedic treatises, and his lyric and philosophical poetry, Dante's comprehensive approach to the sciences of the human reveals the varied perspectives he contributed to the intellectual history of his time.

Notes

1. *Inf*. 3.55–57: "So long a file / of people—I should never have believe / that death could have unmade so many souls." Here and throughout this chapter, I cite from Anna Maria Chiavacci Leonardi's commented edition (Alighieri 1991–94), with Allen Mandelbaum's English translation (Alighieri 1995).
2. *Purg*. 2.79–81: "O *shades*—in all except appearance—*empty*! / Three times I clasped my hands behind him and / as often brought them back against my chest." *Inf*. 6.34–36: "We walked across the

shades on whom there thuds / that heavy rain, and set our soles upon / their *empty images* that seem like persons." My emphases.

3. The philosophical and scientific definition of the souls in the *Commedia* is at the core of Gragnolati (2008), that contextualizes Dante's innovations in the intellectual history of medieval Italian cultures. On bodies and corporeal identity in Dante, see also Jacoff (2000).
4. *Purg*. 3.16–18: "Behind my back the sun was flaming red; / but there, ahead of me, its light was shattered / because its rays were resting on my body."
5. Barolini (2014) speaks for these first purgatorial cantos of "an ongoing meditation on the body," that puts physical consistencies in communication with an epistemological meditation on human limits to knowing. On bodily consistency between *Purgatorio* and *Paradiso*, see also Ureni (2010), and Jacoff (2000).
6. *Purg*. 3.22–26: "and he, my only comfort, as he turned / around, began: "Why must you still mistrust? / Don't you believe that I am with—and guide—you? / The body from within which I cast shadows / is buried where it now is evening…".
7. *Purg*. 3.31–33: "The Power has disposed such bodiless / bodies to suffer torments, heat and cold; / how this is done, He would not have us know."
8. *Purg*. 24.49–57: "But tell me if the man whom I see here / is he who brought the new rhymes forth, beginning: / 'Ladies who have intelligence of love.'" / I answered: "I am one who, when Love breathes / in me, takes note; what he, within, dictates, / I, in that way, without, would speak and shape." / "O brother, now I see," he said, "the knot / that kept the Notary, Guittone, and me / short of the sweet new manner that I hear."
9. *Purg*. 3.106–09: "I turned to look at him attentively: / he was fair-haired and handsome and his aspect / was noble—but one eyebrow had been cleft / by a swordstroke."
10. Freccero references Erich Auerbach (2003: 3–23), whose essay on Odysseus's scar underlines the modes of representation of violence and male identity into the body of the protagonist.
11. In a Dantean perspective, see Boyde (1981), and the bibliography cited in the course of this chapter.
12. Dante's enrollment among doctors and apothecaries attracted a lot of attention from historians of medicine at the beginning of the last

century. A series of notes in the *British Medical Journal* not only reconstructed the social status of the guild in medieval Florence, but also supposed that Dante's interest in medical professionals might have been sparked by his connection with Beatrice Portinari, whose father Folco founded the Spedale di San Matteo—later di Santa Maria Nuova—, a charitable institution providing healthcare to the poor of the city since around 1285 (BMJ 1910: 1, 331; BMJ 1910: 1, 395). However, as both Tonelli and Inglese remarks, Dante's choice of a guild is likely to be a result of his philosophical studies, on account of the curricular proximity between the two disciplines (Inglese 2015: 15; Tonelli 2021: 227).

13. Philosophy nurtured the development of medicine during the thirteenth century, especially in Italy, through a deep study of scientific texts that constituted the backbone of the education in the arts (Siraisi 1981, 1987, 1990; Agrimi and Crisciani 1988). As Tonelli notes (2021: 228), recent scholarship has underlined the web of relations between philosophy, medicine, and literature, not only from the material standpoint of textual transmission (Chandelier 2017), but also from the social viewpoint of personal connections and intellectual horizons (Fenzi 1999; Gentili 2005; Tonelli 2015; Casagrande and Fioravanti 2016).
14. Fioravanti (2014: 15–18) reminds us that if Dante was indeed in Bologna during the years 1304–06 (Indizio 2004, 2005; Gorni 2008), the physicians Dino del Garbo and Bartolomeo di Varignana were teaching in the *Studium* as well. In 1297, Dino was listed as a member of the same guild in which Dante enrolled in 1295 (Siraisi 1981: 57), while Bartolomeo was "a prominent member of the White Guelf faction," thus in the political vicinities of the exiled Florentines of which Dante was himself a part (Siraisi 1981: 47–48). For a thorough appraisal of chronology and geography of *Convivio* and *De vulgari eloquentia*, see Inglese (2015: 72–74, 81–83). For the Bolognese medical school at the time of Dante, see Duranti (2022).
15. Guido da Pisa, *Expositiones et Glose super Comediam Dantis*, *Inf*. 4.143: "Galienus fuit de Grecia, et intantum etiam dum viveret notus in mundo, quod Rome insonuit fama eius. [...] Nam, sicut cum dicimus «prophetam» intelligimus David per anthonomasiam, idest per excellentiam, et cum dicimus «apostolum» intelligimus Paulum, et cum dicimus «evangelistam» intelligimus Iohannem;

et cum dicimus «philosophum» intelligimus Aristotilem; et cum dicimus «poetam» intelligimus Virgilium; ita, cum Avicenna dicit «medicum», intelligit Galienum. Fecit autem, ut fertur, a medicis in ipsa scientia medicine plus quam C volumina, de quibus habentur fere volumina sexaginta, quorum nomina omitto propter tedia evitanda" [Galen was from Greece, and during his lifetime, he was so famous worldwide that his fame resounded in Rome. In fact, just as we say "the Prophet" and mean David by antonomasia, i.e., for excellence, and as we say "the Apostle" and mean Paul, and as we say "the Evangelist" and mean John, and as we say "the Philosopher" and mean Aristotle, and as we say "the Poet" and mean Virgil, so, when Avicenna says "the Physician," he means Galen. According to physicians, he wrote more than a hundred volumes on the art of medicine, of which sixty have been preserved. I omit their titles to avoid causing boredom]. Guido also quotes from Galen when treating Socrates' death, discussed both in *Inf*. 4.134 and *Inf*. 8.19. Commentary accessed through the Dartmouth Dante Project, https://dante.dartmouth.edu/. Accessed 8 July 2024, my translations here and below. On medicine in the *Commedia* and the early commentaries, see also Prioreschi (1994).

16. Benvenuti de Rambaldis de Imola, *Comentum super Dantis Aldigherij Comoediam*, *Par.* 9.57: "Chi'l pesasse ad oncia ad oncia: ut verum dicit; nam homo inter animalia habet multum sanguinis; unde Galenus scribit de uno cui per nares effluxerunt viginti quinque librae sanguinis, et mortuus est" [He is right: in fact, men among all living creatures have a lot of blood. Hence Galen wrote about a man whose nose bled twenty-five pounds of blood, and died]. Commentary accessed through the Dartmouth Dante Project, https://dante.dartmouth.edu/. Accessed 8 July 2024, my translations here and below.
17. *Comentum*, *Par.* 33.94–96: "Un punto solo, idest, una minima particula rerum divinarum, m'è maggior letargo, idest, maior infirmitas memoriae; est enim letargum, ut tradunt Hippocras, Galienus, Avicenna et alii physici, oppressio cerebri cum oblivione et continuo somno, quasi dicat: plus me sopit et smemorat, che venticinque secoli, idest, viginti quinque centenaria annorum; tot enim transcurrerant postquam facta est illa magna navis; all'impresa che fe Nettuno ammirar l'ombra d'Argo, idest, prima navis. Et hic nota quomodo ista comparatio propriissime

exprimit admirationem autoris. Vult enim dicere quod fuit ita stupefactus, visa illa mirabili forma nova oculis suis, sicut Neptunus olim quando vidit mare navigare regnum suum, quia non poterat imaginari quo ingenio, subtilitate vel arte esset facta talis fabrica, et per mare incolumis duceretur: vel non poterat satis mirari et adhuc miratur post tot millia annorum, quo homines exposuerunt animas suas tam manifesto periculo, sicut clamant omnes autores, et pulcre Seneca libro tragoediarum" [*Un punto solo*, i.e., a small particle of divine things, *m'è maggior letargo*, i.e., a major illness of memory; in fact, as Hippocrates, Galen, Avicenna and other natural philosophers say, [lethargy] is an oppression of the brain with oblivion and continuous sleeping, as if he said: more it puts me to sleep and makes me forget, *che venticinque secoli*, i.e., twenty-five centuries; so many in fact passed after that first great ship had been made; *all'impresa che fe Nettuno ammirar l'ombra d'Argo*, i.e., the first ship. And here note how this comparison is most proper to express the admiration of the author. In fact, he wants to say that he was so stupefied, at the sight of such wondrous form, as Neptune once was when he saw his sea kingdom navigated, because he could not fathom how through intelligence, subtlety or art that construction had been made, through which men put their souls to such a great risk, as all the authors claim, and beautifully Seneca in his tragedies]. On the intersection between scientific and mystical discourses in Dante's dreams, see also Bartoli and Ureni (2004).

18. For further medical and philosophical discussions in Benvenuto's commentary and Dante's Ulysses, see Bartoli (2012), Gentili (2017, 2021).
19. It should be noted that the image of *Paradiso* 26 is much more felicitous than the discussion on vision in *Conv.* 2.9.4–5 and 3.9.7–4, for the immediacy of the lexical choices. As a matter of fact, the "gonna" now translates the calque "tunica della pupilla" (*Conv.* 3.9.13). See Bertini Malgarini (1989: 70n).
20. A point that, in the present volume, Eva Del Soldato develops with respect to Renaissance readings of Dante on lovesickness (Chapter 7).
21. In these brief introductory remarks, I wish to recognize my debts towards Alessandro Di Nuzzo's recent anthology of physician

dantisti (2021), who engaged with the *Commedia* from their clinical experience, thus showing the solidity of a tradition that goes back to the fourteenth-century commentaries on the poem.

22. See Cesare Lombroso's "La nevrosi in Dante e Michelangelo [1983]" (Di Nuzzo 2021: 123–28); Bernardo Chiara's "Dante e la psichiatria. Lettera a Cesare Lombroso [1984]" (129–48).
23. See, for instance, Arturo Castiglioni's "La medicina ai tempi e nell'opera di Dante [1922]" (Di Nuzzo 2021: 221–54); Liborio Giuffré's "Il sangue, il cuore e la circolazione," "La funzione della riproduzione," "La dottrina dell'anima," "Le emozioni" (55–86); Antonino Del Gaudio's "Malebolge. La 'febbre acuta' dei falsari della verità," "Lo scheletro, il temperamento, le malattie di Dante," "Morte per perniciosa malarica" (87–112); and Libertino Alajmo's "Dante soffrì veramente di malattie agli occhi? [1922]" (149–62).
24. Not to mention Nardi's exchanges with Étienne Gilson, whose work (1939) uncovered some of the most technical aspects of Dante's thought and his "practical" approach to philosophy. For a recent assessment of Nardi's approach to medieval philosophy within the intellectual contexts of the nineteenth and twentieth centuries, see Falzone and Lenzi (2024).
25. It is not possible to give a full account of Bruno Nardi's contribution to the study of medieval philosophy, but for what concerns Dante and medicine, see at least: (1960), for thorough discussions on the origin of the soul and medical psychosomatic debates; (1966), for the pivotal "L'amore e i medici medievali" (238–67), on therapy against lovesickness and its background in medieval Galenism; (1967), with the essay on Dante and Pietro d'Abano (40–62), and on the stages of life as expounded in the *Convivio* (110–38); and (1983), especially for the essay "Filosofia dell'amore nei rimatori italiani del Duecento e in Dante" (1–92), which also deals with the clashes between Aristotelian cardio-centrism and the different approaches of medieval Galenism. For an updated bio-bibliographical study on Nardi, with special attention to his studies on Dante, see Falzone (2013), Falzone and Lenzi (2024).
26. Fundamental starting points on Dante's medical language are Contini (1970), Baldelli (1978), Bertini Malgarini (1989).
27. On several of these questions, see Bartoli and Ureni (2003, 2004, 2005, 2006) and Ureni (2010, 2014a, 2014b, 2015, 2018).

References

Agrimi, Jole, and Chiara Crisciani. 1988. *Edocere medicos. Medicina scolastica nei secoli XIII–XV*. Naples: Guerini.

Alighieri, Dante. 1991–94. *La Divina Commedia*, ed. Anna Maria Chiavacci, 3 vols. Milan: Mondadori.

Alighieri, Dante. 1995. *The Divine Comedy*. Trans. Allen Mandelbaum. New York: Alfred A. Knopf.

Antonelli, Armando. 2010. Rime estravaganti di Dante provenienti dall'Archivio di Stato di Bologna (con un approfondimento di ricerca sul sonetto della Garisenda vergato da Enrichetto delle Querce). In *Le rime di Dante*, ed. Claudia Berra and Paolo Borsa, 83–116. Milan: Cisalpino Istituto Editoriale Universitario.

Antonelli, Armando. 2018. Aggiunte al corpus delle rime dei Memoriali bolognesi. *Medioevo Letterario D'italia* 15: 167–79.

Antonelli, Armando, and Riccardo Pedrini. 2001. Appunti sulla più antica attestazione dell'*Inferno*. *Studi e problemi di critica testuale* 63: 29–41.

Ardizzone, Maria Luisa. 2011. *Dante: il paradigma intellettuale. Un'inventio degli anni fiorentini*. Florence: Olschki.

Ardizzone, Maria Luisa. 2016. *Reading as the Angels Read: Speculation and Politics in Dante's Banquet*. Toronto: Toronto University Press.

Auerbach, Erich. 2003. *Mimesis: The Representation of Reality in Western Literature*. Trans. William R. Trask. Princeton: Princeton University Press.

Baldelli, Ignazio. 1978. Lingua e stile nelle opere volgari di Dante. In *Enciclopedia Dantesca*, vol. 6 (*Appendice*), 55–112. Rome: Istituto dell'Enciclopedia Italiana.

Barolini, Teodolinda. 1992. *The Undivine Comedy: Detheologizing Dante*. Princeton: Princeton University Press.

Barolini, Teodolinda. 2014. *Purgatorio* 3: Blond, and Beautiful. In *Commento Baroliniano*, Digital Dante. New York: Columbia University Libraries. https://digitaldante.columbia.edu/dante/divine-comedy/purgatorio/purgatorio-3/. Accessed on 21 July 2023.

Bartoli, Vittorio. 2012. La complessione calda di Ulisse. Chiosa al commento di Benvenuto da Imola a *Inf*. XXVI. *Studi danteschi* 77: 105–14.

Bartoli, Vittorio, and Paola Ureni. 2003. Controversie medico-biologiche in tema di generazione umana nel XXV del *Purgatorio*. *Studi Danteschi* 48: 83–111.

Bartoli, Vittorio, and Paola Ureni. 2004. Sonno e 'animi deliquium' nel viaggio ultraterreno di Dante. *Studi Danteschi* 49: 211–29.

Bartoli, Vittorio, and Paola Ureni. 2005. La dottrina di Galeno in 'sangue perfetto.' *Studi Danteschi* 70: 335–43.

Bartoli, Vittorio, and Paola Ureni. 2006. La morte cruenta di Jacopo del Cassero e di Bonconte da Montefeltro (*Purg*. V 73–102). Una nuova lettura fondata sulla scienza medica medievale. *Studi Danteschi* 71: 9–26.

Bellomo, Saverio. 2004. Primi cultori di Dante a Bologna. In *Bologna nel Medioevo: atti del convegno (Bologna, 28–29 ottobre 2002)*, 207–22. Bologna: Pàtron.

Bertini Malgarini, Patrizia. 1989. Il linguaggio medico e anatomico nelle opere di Dante. *Studi danteschi* 61: 29–108.

Borsa, Paolo. 2007. *La nuova poesia di Guido Guinizelli*. Fiesole: Cadmo.

Boyde, Patrick. 1981. *Dante Philomythes and Philosopher: Man in the Cosmos*. Cambridge: Cambridge University Press.

British Medical Journal. 1910a. Was Dante a Doctor? (1). February 5.

British Medical Journal. 1910b. Was Dante a Doctor? (2). February 12.

Casagrande, Carla, and Gianfranco Fioravanti, eds. 2016. *La filosofia in Italia al tempo di Dante*. Bologna: il Mulino.

Chandelier, Joël. 2017. *Avicenne et la médecine en Italie. Le Canon dans les universités italiennes (1200–1350)*. Paris: Honoré Champion.

Contini, Gianfranco. 1970. Sul XXX dell'*Inferno*. In *Varianti e altra linguistica*, 447–58. Turin: Einaudi.

Corti, Maria. 1982. *Dante a un nuovo crocevia*. Florence: Le lettere.

Corti, Maria. 1983. *La felicità mentale: Nuove prospettive per Cavalcanti e Dante*. Turin: Einaudi.

Di Nuzzo, Alessandro, ed. 2021. *Dante e la medicina. L'arte medica e farmaceutica nell'opera dantesca*. Reggio Emilia: Aliberti.

Duranti, Tommaso. 2022. Un nuovo sapere sull'uomo: la scuola medica bolognese ai tempi di Dante. In *Dante e Bologna. Istituzioni, convergenze e saperi*, ed. Armando Antonelli, and Franziska Meier, 181–94. Ravenna: Giorgio Pozzi Editore.

Falzone, Paolo. 2013. Bruno Nardi's Louvain Dissertation (1911) and the Uneasy Character of Dante's Philosophy. *Tijdschrift Voor Filosofie* 75: 357–73.

Falzone, Paolo, and Massimiliano Lenzi. 2024. *Tra Rosmini e Gentile. Il Medioevo teoretico di Bruno Nardi*. Florence: SISMEL-Edizioni del Galluzzo.

Fenzi, Enrico. 1999. *La canzone d'amore di Guido Cavalcanti e i suoi antichi commenti*. Genova: il melangolo.

Fioravanti, Gianfranco. 2014. Introduzione. In Dante Alighieri, *Opere*, dir. Marco Santagata, vol. 2: *Convivio, Monarchia, Epistole, Egloge*, 5–79. Milan: Mondadori.

Freccero, John. 1986. *Dante: The Poetics of Conversion*, ed. Rachel Jacoff. Cambridge, MA: Harvard University Press.

Gentili, Sonia. 2004. Destini incrociati. Taddeo Alderotti docente allo studio bolognese e la letteratura volgare delle origini. In *Bologna nel Medioevo: atti del convegno (Bologna, 28–29 ottobre 2002)*, 165–206. Bologna: Pàtron.

Gentili, Sonia. 2005. *L'uomo aristotelico alle origini alle origini della letteratura italiana*. Rome: Carocci.

Gentili, Sonia. 2006. L'*Etica volgarizzata* da Taddeo Alderotti (m. 1295): Saggio e commento. *Documenti e Studi Sulla Tradizione Filosofica Medievale* 17: 249–81.

Gentili, Sonia. 2012a. La vulgarisation de l'*Éthique d'Aristote* en Italie aux XIIIe et XIVe siècles: enjeux littéraires et philosophiques. *Médiévales* 63: 47–58.

Gentili, Sonia. 2012b. Amicizia, città e spazio sociale nell'*Etica di Aristotele* volgarizzata da Taddeo Alderotti. In *Parole e realtà dell'amicizia medievale: atti del convegno di studio svoltosi in occasione della XXII edizione del Premio internazionale Ascoli Piceno (Ascoli Piceno, Palazzo dei Capitani, 2–4 dicembre 2010)*, ed. Isa Lori Sanfilippo, and Antonio Rigon, 127–44. Rome: Istituto storico italiano per il Medio Evo.

Gentili, Sonia. 2014. L'edizione dell'*Etica in volgare* attribuita a Taddeo Alderotti: risultati e problemi aperti. In *Aristotele fatto volgare: tradizione aristotelica e cultura volgare nel Rinascimento*, ed. David A. Lines, and Eugenio Refini, 39–59. Pisa: ETS.

Gentili, Sonia. 2017. Novità su Dante e Omero: il fuoco di Ulisse. In *I classici di Dante*, ed. Paola Allegretti, and Marcello Ciccuto, 11–24. Florence: Le Lettere.

Gentili, Sonia. 2021. L'Ulisse omerico da Aristotele a Dante. In *Dante e la tradizione classica: atti del convegno in ricordo di Saverio Bellomo (Pisa, Scuola Normale Superiore, 10–11 aprile 2019)*, ed. Claudia Russo and Stefano Carrai, 107–16. Ravenna: Longo.

Gilson, Étienne. 1939. *Dante et la philosophie*. Paris: Vrin.

Gorni, Guglielmo. 2008. *Dante. Storia di un visionario*. Rome-Bari: Laterza.

Gragnolati, Manuele. 2008. *Experiencing the Afterlife: Soul and Body in Dante and Medieval Culture*. Notre Dame: Notre Dame Press.

Gubbini, Gaia, ed. 2020. *Body and Spirits in the Middle Ages: Literature, Philosophy, Medicine*. Berlin: De Gruyter.

Indizio, Giuseppe. 2004. Le tappe venete dell'esilio di Dante. *Miscellanea Marciana* 19: 35–64.

Indizio, Giuseppe. 2005. Dante secondo i suoi antichi (e moderni) biografi: saggio per un nuovo canone dantesco. *Studi Danteschi* 70: 237–94.

Inglese, Giorgio. 2015. *Vita di Dante. Una biografia possibile*. Rome: Carocci.

Jacoff, Rachel. 2000. "Our Bodies, Our Selves": The Body in the *Commedia*. In *Sparks and Seeds: Medieval Literature and Its Afterlife. Essays in Honor of John Freccero*, eds. Dana E. Stewart and Alison Cornish, 119–37. Turnhout: Brepols.

Kristeller, Paul Oskar. 1955. A philosophical treatise from Bologna dedicated to Guido Cavalcanti: magister Jacobus de Pistorio and his "Questio de felicitate." In *Medioevo e Rinascimento. Studi in onore di Bruno Nardi*, 425–63. Florence: Sansoni.

Malatesta, Maria. 2005. Il romanzo: testimonianza e rappresentazione. *Contemporanea* 8 (4): 698–703.

Nardi, Bruno. 1911. Sigieri di Brabante e le fonti della filosofia di Dante. *Rivista Di Filosofia Neo-Scolastica* 3: 187–95, 526–45.

Nardi, Bruno. 1912. Sigieri di Brabante e le fonti della filosofia di Dante. *Rivista Di Filosofia Neo-Scolastica* 4: 73–90, 225–39.

Nardi, Bruno. 1922. Raffronti fra alcuni luoghi di Alberto Magno e di Dante. *Giornale Storico Della Letteratura Italiana* 80: 295–304.

Nardi, Bruno. 1949. Dante profeta. In *Dante e la cultura medievale*, ed. Paolo Mazzantini, 336–416. Bari: Laterza.

Nardi, Bruno. 1960. *Studi di filosofia medievale*. Rome: Edizioni di storia e letteratura.

Nardi, Bruno. 1966. *Saggi e note di critica dantesca*. Milan: Ricciardi.

Nardi, Bruno. 1967. *Saggi di filosofia dantesca*. Florence: La Nuova Italia.

Nardi, Bruno. 1983. *Dante e la cultura medievale*, ed. Paolo Mazzantini. Bari: Laterza.

Orlando, Sandro. 1981. *Rime dei Memoriali bolognesi, 1279–1300*. Turin: Einaudi.

Pace, Matteo. 2022. "Come vertute in petra prezïosa": An Avicennian Thread in Guinizzelli's Theory of Nobility. *Romanic Review* 113 (1): 65–86.

Prioreschi, Plinio. 1994. Medicine in the *Divine Comedy* and Early Commentaries. *The Journal of Medical Humanities* 15 (1): 51–72.

Siraisi, Nancy. 1981. *Taddeo Alderotti and His Pupils: Two Generations of Italian Medical Learning*. Princeton: Princeton University Press.

Siraisi, Nancy. 1987. *Avicenna in Renaissance Italy*. Princeton: Princeton University Press.

Siraisi, Nancy. 1990. *Medieval and Early Renaissance Medicine: An Introduction to Knowledge and Practice*. Chicago: University of Chicago Press.

Tavoni, Mirko. 2011. Introduzione. In Dante Alighieri, *Opere*, dir. Marco Santagata, vol. 1: *Rime, Vita Nova, De Vulgari Eloquentia*, 1067–116. Milan: Mondadori.

Tonelli, Natascia. 2015. *La fisiologia della passione. Poesia d'amore e medicina da Cavalcanti a Boccaccio*. Florence: SISMEL-Edizioni del Galluzzo.

Tonelli, Natascia. 2021. Medicine. In *The Oxford Handbook of Dante*, ed. Manuele Gragnolati, Elena Lombardi, and Francesca Southerden, 227–41. Oxford: Oxford University Press.

Ureni, Paola. 2010. Human Generation, Poetic Creation and Memory: From the *Purgatorio* to the *Paradiso*. *Quaderni D'italianistica* 21 (2): 9–35.

Ureni, Paola. 2014a. Aspetti eterodossi nel pensiero medico. Una filigrana galenica nella scrittura di Dante. In *Ortodossia ed eterodossia in Dante Alighieri. Atti del convegno di Madrid (5–6 novembre 2012)*, ed. Carlota

Cattermole, Celia de Aldama, and Chiara Giordano, 303–24. Alpedrete: Ediciones de La Discreta.

Ureni, Paola. 2014b. Medicine and Radical Thought: A Possible Galenic Presence in the *Commedia*. In *Dante and Heterodoxy: The Temptations of 13th-Century Radical Thought*, ed. Teodolinda Barolini and Maria Luisa Ardizzone, 225–41. Newcastle upon Tyne: Cambridge Scholars Publishing.

Ureni, Paola. 2015. Lo 'stupor mentis' a occhi aperti di Vanni Fucci (*If*. XXIV 97–118). Filigrane mediche e scritturali. *Dante. Rivista Internazionale Di Studi Su Dante Alighieri* 12: 55–67.

Ureni, Paola. 2018. Medicine and Dante's Political Thought. In *Dante as Political Theorist: Reading Monarchia*, ed. Maria Luisa Ardizzone, 209–22. Newcastle upon Tyne: Cambridge Scholars Publishing.

van Straten, Giorgio. 2023. *Invasione di campo. Quando la letteratura racconta la storia*. Milan: Laterza.

Zavattero, Irene. 2005. La "Quaestio de felicitate" di Giacomo da Pistoia. Un tentativo di interpretazione alla luce di una nuova edizione critica del testo. In *Le felicità nel Medioevo*, ed. Maria Bettetino and Francesco D. Paparella, 355–409. Louvain-la-Neuve: Fédération Internationale des Instituts d'Études Médiévales.

CHAPTER 2

When the Ocean Towered Above the Earth: Dante's *Questio* in Context

Ivano Dal Prete

The last few years have seen a certain return of interest toward Dante Alighieri's most properly scientific work, the controversial *Questio de aqua et terra (On the Problem of Water and Earth)*.[1] Allegedly discussed in a church of Verona in January 1320, the *Questio* engages with one of the most debated meteorological problems of his time, namely the natural reasons for the existence of dry land, proposing that a "hump" in the otherwise spherical earth allows part of it to emerge from the ocean. After Alison Cornish's seminal study of Dante's meteorology (2004), as of late scholars such as Paolo Pellegrini (2021), Luna Sarti (2019), and Gianfranco Fioravanti (2019) have examined various facets of Dante's *Questio* or of closely related subjects. Among them, Fioravanti has urged scholars to move on from the lengthy dispute on the authenticity of the lecture, and to pay instead more attention to the larger philosophical and cultural

I. Dal Prete (✉)
Department of History, Yale University, New Haven, CT, USA
e-mail: ivano.dalprete@yale.edu

M. Pace (ed.), *Dante and the Sciences of the Human*, Palgrave Studies in Medieval and Early Modern Medicine,
https://doi.org/10.1007/978-3-031-69253-6_2

context in which it was produced (2019: 82). Regardless of the authorship, the *Questio* remains indeed a remarkable and valuable source on one of the most debated issues in medieval (and Renaissance) science—as well as on Dante's own scientific culture, if the attribution supported by most modern scholars is correct.[2]

In this chapter I take a further step in that direction, focusing not so much on the possible direct or indirect sources of the *Questio*—learned and comprehensive surveys are already available, albeit often dated (Boffito 1902; Duhem 1958; Alighieri 1979; Fioravanti 2019)—but rather on the general historical and scientific framework of the problem and on the often misunderstood theory Dante vehemently opposes: namely that the surface of the "sphere of water" would be in part of its circumference higher than dry land, and in particular, that the center of the sphere of water would not coincide with that of the universe and of the sphere of earth. The author of the *Questio* and modern scholars alike have regarded this construct as absurdly illogical since it seems to defy plain common sense. On the contrary, I suggest that this solution gave plausible answers to difficult physical problems other models (including Dante's favorite) struggled to address, and that in Dante's time that doctrine was well known and widely accepted even in didactic literature and in vernacular texts. Far from an example of failed reasoning and logical fallacies, it exemplifies the sophistication of medieval minds and their capacity to entertain explanations that transcended the immediate perception of the senses.

The Problem of Water and Earth

> The question then concerned the location and shape, or form, or the two elements, Water namely and Earth... And the question was confined to this – as the very foundation of the truth to be investigated – that the inquiry should be: whether water in its own sphere, that is its natural circumference, is anywhere higher than the earth which emerges from the waters and which we commonly call "the habitable quarter." (Alighieri 1903: 4–7)[3]

It would not be an exaggeration to state that the issue so neatly framed in the opening paragraphs of the *Questio* represents *the* fundamental problem of medieval meteorology. The relationship and relative positions of earth and water in the terraqueous globe affected its subterranean and superficial geography; the cycle of water and the origin of fountains; and even whether the earth as a whole remained at rest in the center of the universe or was involved instead in some sort of global movement. Because of its vast implications, it was discussed not only in proper meteorological texts but even in commentaries on Aristotle's *On the Heavens* and the *Physics*, as well as in a plethora of encyclopedias both Latin and vernacular.

The problem at the heart of the controversy was the hard fact that dry land does exist.[4] According to natural principles as medieval philosophers knew them, this should not have been the case: earth, water, air, and fire were expected to settle in this order around the center of the universe in concentric spheres, following their specific weights and their tendency to move toward their natural place. A sphere of earth completely surrounded by a lighter sphere of water would then be the ideal state of equilibrium of the world, with each element in its natural place—or as close as possible to it. Of course, experience showed otherwise. The "why" (the final cause) seemed easy enough: dry last must exist in order to support the generation of terrestrial animals, including human beings as part of God's providential plan for the universe. How it managed to do so (the efficient cause) was far less straightforward. For scholars such as William of Auvergne, Michael Scot, or Campanus of Novara, direct divine intervention was the necessary efficient cause: as a permanent miracle, it lay outside the scope of natural philosophy and needed no further inquiry. Still, most authors maintained that God acted in the universe through natural causes, and that miracles should not be invoked if plausible natural reasons could be found.

Before the reintroduction of Aristotle's natural philosophy in the course of the twelfth and thirteenth century, the problem was often framed within a physical history of the universe largely based on Plato's *Timaeus*, in which the world evolves from an initial to a final state. According to the author of the anonymous treatise *On the Elements*, composed probably in Sicily around 1160, God in the beginning created the elements whose motion gave origin to the material universe as we see it. Their "atoms" (defined as the smallest particles of an element) move according to their nature, fire being the fastest while the earth lies almost motionless at the center of the world surrounded by water and air (Dales 1965; Dales 1990). In the roughly coeval *On the Philosophy of the World*, William of Conches showed that the terraqueous globe had reached its present form through natural processes that carried out the Divine plan for the Creation (William of Conches 1531: 16). William noted that when philosophers mentioned the Creation of the world, they did not mean its celestial or terrestrial bodies but only the four elements. Their motion then gave origin to everything else, without any need for further supernatural intervention.[5] Following their natural inclination, earth, water, air, and fire fell therefore toward the center of gravity of the universe, where they occupied four concentric spheres according to their specific weight. The skies were made of the same elements found on the Earth, though in different proportions.

In the context of a universe with a directional evolution, the emergence of dry land could be physically explained through the vapors raised by the Sun, whose heat continuously turned water into air and fire causing a progressive diminution of the sea level until part of the element of earth remained uncovered. In some versions, the process was still underway and would continue until the complete desiccation of the globe and the extinction of life on Earth.[6] Far from contradicting the Genesis creation story, such accounts provided a physical counterpoint that supported its truth. While Scripture taught the "why", recurring to images and metaphors adapted to the understanding of common people, natural philosophy investigated the actual "how". When correctly interpreted, neither contradicted the other. In both cases, the Earth was shown to evolve from a disordered beginning to a state that made terrestrial life and the existence of human beings possible: this condition was not stable though, as the physical and moral decadence of the world inexorably led toward its destruction or regeneration. The emersion of part of the element of earth caused by the evaporation of the primeval ocean corresponded to God's command that separated water from dry land in Genesis; its final desiccation, to the "great fire" that will bring the end of the world.

Such models could not be applied readily to the Aristotelian Earth that was as eternal and fundamentally immutable as the rest of his universe. Change could only be cyclical, including in the lower world we inhabit: the fundamental task of meteorology as a scholarly discipline consisted in discerning the patterns of causality and predictability that could be glimpsed through the apparent chaos of the earthly phenomena (Ducos 1998; Martin 2011). Mountains were ruthlessly leveled by erosion, but different ones rose; while old rivers dried up, new ones sprang; great floods and droughts periodically drowned or scorched whole countries;

as one side of the continent became slowly engulfed by the ocean, new dry land emerged somewhere else. Taken as a whole, the Earth looked always the same.

There were more problems. Agitated by the motions, heat, and light of the celestial bodies, the four elements keep changing into one another thus causing the processes of generation and corruption. However, the stability of Aristotle's universe required that their respective quantity must always be roughly the same. Water is lighter than earth though, meaning that it must occupy a larger volume for a given quantity. Aristotle supposed that it may take ten measures of air to make one measure of water, and vice versa; many commentators inferred that the proportion between water and earth, air and water, and air and fire should then be of the same order (Duhem 1958: 9.91–96). Indeed, it was a common tenet of medieval meteorology (albeit by no means unquestioned!) that the volume of the sphere of water could be up to ten times larger than that of the earth, and its radius up to twice as much. If so, water could not be a thin layer that gathered in the cavities of the earth's surface or the underground. Finally, every model based on a static Earth would need to solve not only the problem of how dry land came to emerge but also of how it could be maintained in the face of continued erosion: the unrelenting action of waters and winds kept washing away the earth to the ocean, until every height would be leveled and water would once again reclaim the whole of the globe's surface.

Of course, no one in Dante's time could claim that the Earth was actually eternal if not "philosophically speaking." A universe coeternal with God could still be entertained in the twelfth century, when Bernard Silvestris's *Cosmographia* described the material world as going through never-ending cycles because its cause was eternal too (Dales 1990: 34; Stock 2015: 11).[7] This avenue closed in 1215, when the fourth Lateran Council established the dogma that the world did have a beginning (Tanner 1990: 2.230; Dal Prete 2022: 30). Still, nothing was determined as to when and how the Creation took place. As I have shown elsewhere, medieval culture did not feel constrained by the short chronology of the Bible and many were willing to admit the temporal scales needed by erosion to carry out its effects (Dal Prete 2022: 15–92). In the late 1200s, Giles of Lessines and others cited the very existence of mountains as evidence that the Earth could not be immensely old (Nothaft 2014a,

2014b). In the following decades though, it became increasingly important to show that dry land could be sustained indefinitely, and to look for plausible natural causes that could counter the action of erosion.

Albertus Magnus's discussion of the emersion of dry land in his *Meteorology*, composed toward the middle of the thirteenth century, and a crucial source for Dante (Cornish 2004), was arguably instrumental in the transition from directional conceptions of the Earth's history to the steady-state Aristotelian model. The German Dominican did describe the doctrine of the progressive desiccation of the globe and he seems to have been tempted by that solution, but ultimately, he decided against it. As he argued, the generation of plants and animals must have been going on since the beginning of the world, yet this would be impossible if the earth had once been covered entirely by the waters.[8] His conclusions notwithstanding, the popularity of Albertus Magnus's work ensured that the notion of a directional Earth continued to circulate even in the following centuries. For instance, Leonardo da Vinci's writings on the subject seem to be based on the vernacular version composed in Florence around 1310, of which the Tuscan artist almost certainly owned a copy.[9] In the early fourteenth century though, the problem of the emersion of dry land was normally discussed in the assumption of an Earth that is either static or in a state of dynamic balance. This is also the framework of the *Questio*.

Since the Earth had never been fundamentally different, dry land must have always existed in spite of the apparent instability of this condition. Aristotle solved the conundrum by theorizing that the motion transmitted to the lower world by the celestial bodies prevented the elements from reaching their most stable configuration. In the case of water and earth, the continuous evaporation induced by the Sun's heat was the reason why water did not cover completely the earth. What remained filled the depressions of the surface and the cavities of the underground. As long as the Sun kept shining and rolling along its course, parts of the Earth would always emerge (Boffito 1902: 77–78). Aristotle's solution did not lack backers in Dante's time, like physician Antonio Pelacani and geographer and astronomer Andalò del Negro (Alighieri 1979: 705–06; Duhem 1958: 9.146–48). One century later, Leon Battista Alberti argued likewise that the seas must be just a thin layer sticking to the surface of the Earth, "like spring dew on an apple" (Alberti 1988: 336).[10] Still, the more common assumption that water must occupy a larger (possibly, much larger) volume than earth made this thesis increasingly implausible between the fourteenth and the fifteenth centuries. The *Questio* does not even bother to mention, let alone confute it.

The Hunchback Earth

Dante supports instead the thesis that the sphere of water did in fact encapsulate the earth, yet some of it could pop up from the surface of the global ocean thanks to a providential swelling that made up the inhabited part of the world (Fig. 2.1).

In the late 1200s, Giles of Rome even ventured to reckon the height of the Earth's "hump" at about 5/4th of its radius (Duhem 1958: 9.143–45). It is telling that neither Giles nor Dante contented themselves with figuring that the Earth could simply have been made by God that way (a sufficient answer in a recently created world), but tried to find plausible physical causes that could exert a continuing action. This does not necessarily mean that Dante may have envisaged a world of indefinite duration, but rather that around 1300 the problem was framed in such terms that a "one-off" cause would be considered unsatisfactory. One of the favorite candidates among authors who mentioned the "hump" theory was a sort of magnet-like attraction of the northern stars (considered brighter and more powerful than the southern ones) over the body of the earth that caused and maintained its bulging. This explanation rested on solid authority: Albertus Magnus, whose work proved crucial in the thirteenth-century astrologization of Aristotelian natural philosophy, had

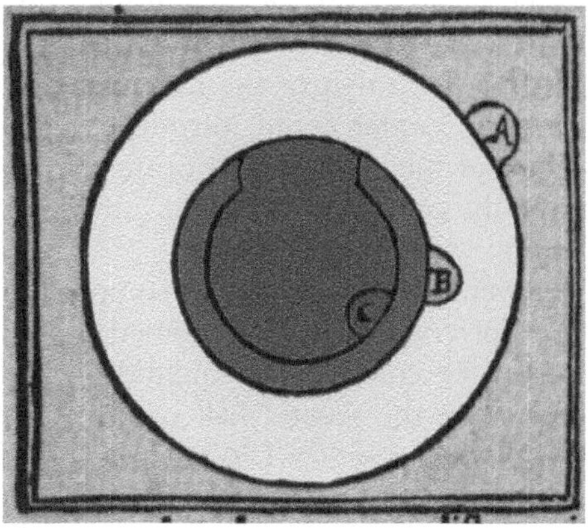

Fig. 2.1 A = sphere of air; B = sphere of water; C = sphere of earth. Colorized from: Dante Alighieri, *Questio florulenta ac perutilis de duobus elementis aquae & terrae tractās nuper reperta que olim Mantuae auspicata*, ed. Giovanni Benedetto Moncetti (Venetiis: per M. de Monteferrato, 1508)

recognized already that of all the elements the earth was extraordinarily receptive to celestial influences (Rutkin 2019: 113–15).[11] Many further assumed that if the swelling of the earth was preserved by the action of certain constellations, then it could not remain fixed in the same place but it must also wander across the globe dragged by their motions—noticeably, the precession of the equinoxes whose period according to Ptolemy amounted to 36,000 years. This was often identified as the movement that caused the cyclical submersion and emersion of parts of the earth mentioned by Aristotle (Dal Prete 2022: 44, 56). In 1271, Robert the Englishman noted that there was "evidence of this in places that used to be underwater and are now dry, as we can see in certain locations in England" (Duhem 1958: 1.133).

The most widespread alternative to the 'hump' theory, and one that was the main target of the author of the *Questio*, held that some earth is dry not because the northern stars raised part of it, but because the sphere of water and that of earth were not concentric: the offset between the centers of the two elements then caused part of the earth to emerge from the larger body of water (Fig. 2.2).

Which one of the two spheres had its center removed from the center of the universe was the object of considerable controversy. Alexander of Aphrodisias argued that because of the great cavities within the earth and its supposedly sponge-like structure, its density was not uniform, and one side will be lighter than the other. While the center of mass of the earth (as we would call it today) coincides with the center of the universe as expected, its geometrical center does not; on the contrary, being a fluid water will always place itself around the center of gravity of the world. In other words, both spheres respected the principle that their centers should coincide with that of the universe, but thanks to the solid nature of the earth Alexander could introduce a distinction between the geometrical center of the earth and its center of mass—"the center of its magnitude" and "the center of its weight" in medieval sources (Duhem 1958: 9.80–81).

Several decades after the composition of the *Questio*, Jean Buridan at the University of Paris combined Alexander's doctrine with the new fascination for concepts of equilibrium and dynamic balance that began to emerge in the second half of the thirteenth century (Kaye 2014: 442–62; Duhem 1958: 9.196–202).[12] Possibly aiming to defuse some of the most frequent objections against the indefinite sustainability of dry land, the famous Master of Arts decoupled the creation of new mountains from

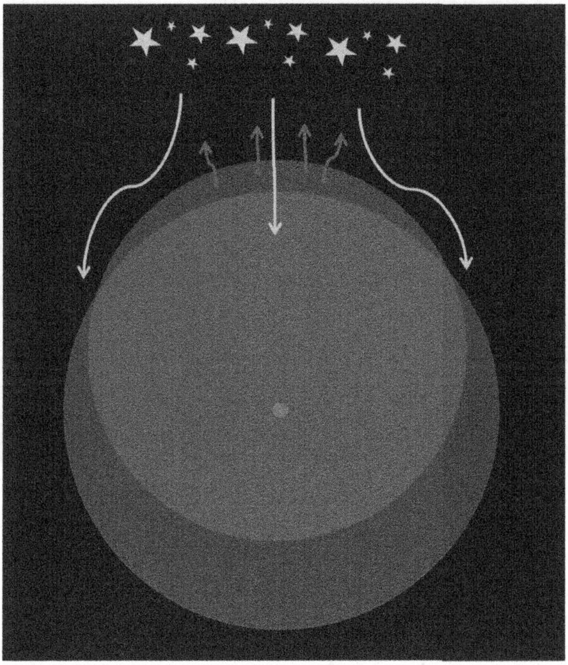

Fig. 2.2 Earth and water according to Restoro of Arezzo

the precession of the equinoxes, whose period was way too fast to account for the observed slowness of geological change. According to Buridan, the exposure to the heat of the Sun and the continuous ablation of earth to the ocean by water erosion tend to make the emerged quarter lighter and lighter. On the contrary, lack of heat and the weight of the ocean pack earth on the opposite side to much higher densities. The center of mass of the earth then constantly shifts away from the emerged side, but since its nature is to be at rest in the center of the universe, the sphere of the earth as a whole must continuously crawl in the direction of the emerged hemisphere to remain in its natural place. As the earth is washed down from the mountains to the sea, where it forms new layers of sediments, what used to be the ocean floor moves toward the center; on the other side new land is pushed up, and so forth ad infinitum. "Nor is there any other way", Buridan concluded, "such mountains can be generated and maintained"[13] (Fig. 2.3).

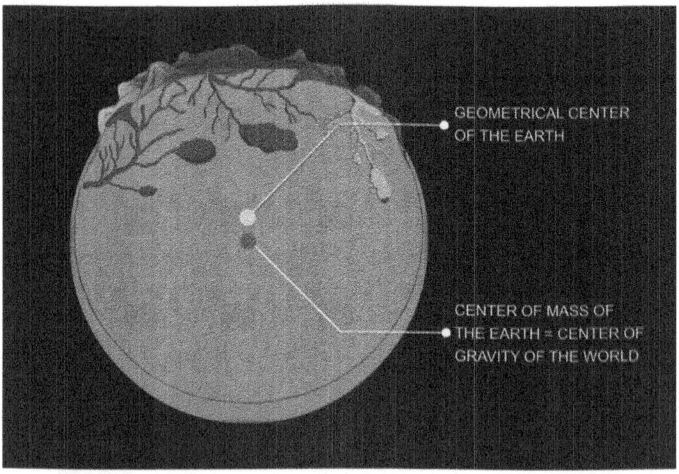

Fig. 2.3 Jean Buridan's model of the Earth

The doctrine of the eccentricity of the earth raised serious questions concerning the quiescence of the earth, but on the other hand, it had welcome theological implications. For instance, while endorsed by a number of authors both ancient and recent (including Aristotle) the existence of the antipodes appeared problematic: since the ocean was considered impassable, humans living in those areas could not be reached by the Christian Revelation and would therefore be excluded from the possibility of salvation. However, if the sphere of earth was eccentric then water must be so deep in the southern hemisphere, that dry land could not possibly emerge there. The eccentricity of the earth required that our "inhabitable quarter" be the only region where some of it can rise above the water—a conclusion Buridan, as usual, tried to buttress with empirical evidence:

> Some have tried to cross the sea to arrive at other quarters, but they have never been able to reach inhabitable land; and so they say that Hercules placed pillars at the limits of the quarter that we inhabit [Gibraltar strait] to signal that beyond them there is no inhabitable land, or seas that can be crossed.[14]

The author of the *Questio* agreed that the earth featured only one inhabitable continent and was everywhere else covered by the ocean, but he flatly rejected its eccentricity. The first of his arguments, namely that the earth is a simple body and therefore it cannot have heavier and lighter parts, appears rather surprising as the presence of underground vapors, waters, fires, exhalations, and cavities seems to have been a given to almost everyone (including Dante himself in the last part of the lecture). While he may have considered such phenomena superficial and unable to affect the earth as a whole, the point is not discussed at all (Alighieri 1979: 756–58). The second reason looks more sensible: geometry shows that an eccentric earth should emerge as a circular dome, yet it was well known that dry land stretched more in longitude than in latitude, its shape resembling a half moon (Alighieri 1979: 762). As to why the northern stars should generate this particular configuration rather than a circular one, Dante quickly brushes away the objection stating that not enough matter was available for a complete circle (Alighieri 1979: 768). How good this explanation might have sounded to his interlocutors in Mantua and to his listeners in Verona is anyone's guess. In any case, the doctrine that incensed the poet and that occasioned his meteorological excursus was not the eccentricity of the earth, but the surprisingly common notion that the sphere of water would be in part of its circumference *higher* (as in, further away from the center of the universe) than even the highest mountains.

A Sea That "Towers Above the Shore"

On the surface, this sounds like utter nonsense. Everyday experience shows that water tends to go down and being a fluid, it will unless it is contained on every side by some solid. Dry land can only exist if it is higher than neighboring seas, rivers or lakes, or if natural or artificial dams prevent them from flooding lower areas—which brings us back to the previous case. Francesco Mazzoni could not help hinting at the apparent lack of common sense of those who argued otherwise (Alighieri 1979: 705). Yet Pierre Duhem acknowledged that this "absurd opinion", as he called it, was "no doubt commonly held in the thirteenth century"—at least among "those with little capacity for scientific reasoning" (1958: 9.128). By all appearances, this was not a fringe theory but a model that seems to have been especially popular in didactic literature; was easily accepted even among vernacular readers, in spite of being highly counterintuitive; and remained current for centuries after Dante supposedly demolished it in his lecture.

The *Questio* discusses and refutes two different ways the sphere of water could be higher than earth: either because it was bulged just like the earth in Dante's favorite theory, but with a larger and taller swelling; or because the sphere of water, rather than that of earth, was eccentric so that the displacement of its center from that of the universe cleared dry land on the opposite side (Figs. 2.4a, b, 2.5).

An essential issue the author of the *Questio* is well aware of, but has been rarely raised by modern scholars, is why did such a doctrine needed to be developed. And, what advantages did it bring over its alternatives? In the following pages, I focus especially on the theory of the eccentricity of water, as it seems to have been the most common and the version that underpinned most of the literature on the subject.

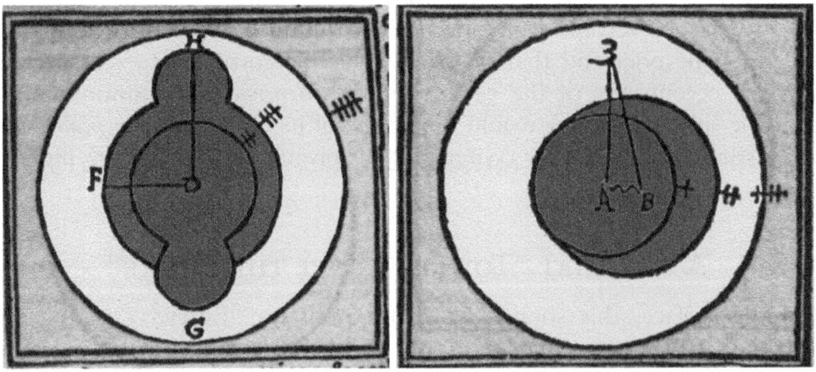

Fig. 2.4 **a** G = swelling of the sphere of earth; H = swelling of the sphere of water; F = surface of the sphere of water. + = sphere of earth; ++ = sphere of water; +++ = sphere of air. Colorized from: Dante Alighieri, *Questio florulenta ac perutilis de duobus elementis aquae & terrae tractãs nuper reperta que olim Mantuae auspicata*, ed. Giovanni Benedetto Moncetti (Venetiis: per M. de Monteferrato, 1508). **b** A = center of the universe = center of the sphere of earth; B = center of the sphere of water; + = sphere of earth; ++ = sphere of water; +++ = sphere of air. Colorized from: Dante Alighieri, *Questio florulenta ac perutilis de duobus elementis aquae & terrae tractãs nuper reperta que olim Mantuae auspicata*, ed. Giovanni Benedetto Moncetti (Venetiis: per M. de Monteferrato, 1508)

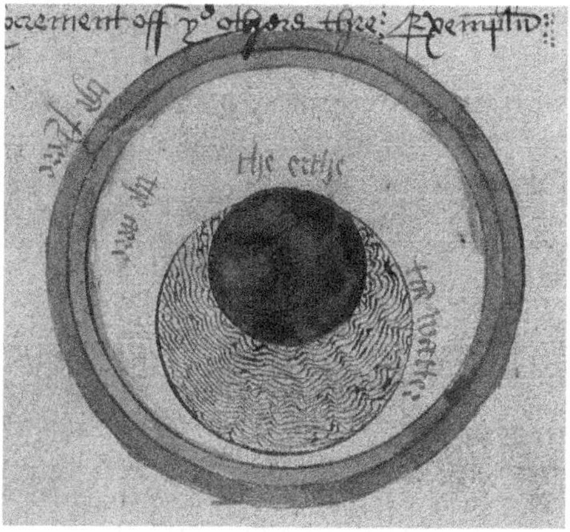

Fig. 2.5 The sphere of earth emerges from a larger sphere of water due to the latter's eccentricity with respect to the center of the universe. Joannes de Sacro Bosco, *De Sphaera, Etc.* Trans. Anthony Askham (Cambridge, 1526), Yale University, Beinecke Library

In his vast (but dated) review of the sources of the *Questio*, Boffito could not find any evidence of its circulation before the fifteenth century: in fact, he used the argument to deny the attribution of the *Questio* to Dante as he could not have addressed doctrines that did not exist yet (1902: 75–88). In his recent reappraisal Fioravanti notices, though, that the theory is present already in Antonio da Parma's commentary on Avicenna's *Canon*, certainly composed before the author died in 1327, and in other slightly posterior writings that all refer to older literature. Without mentioning its origins, in the same decades Buridan called it a "famous opinion" (2019: 85). I would further suggest that while these may be the first known documents referring explicitly to the "eccentricity" of the sphere of water, this doctrine appears to be a definite assumption in important works composed in the previous decades.

Among them, Brunetto Latini's widely popular *Tresor* (c. 1260) and its Tuscan translations might have been crucial for the diffusion into vernacular contexts of the notion that water was higher than earth. Latini not only adopted this solution without any qualms, but he went so far as to cite direct empirical evidence for it, describing how from the shores of "Mauritania" the ocean looked "much higher than the earth, yet it is self-contained so that it neither falls, nor runs over the earth."[15] A modern reader could question the reliability of his source, but this would be beyond the point. What is interesting is that Latini considered the information plausible in the first place; and secondly, that this observation was only credible in the assumption that the sphere of water is eccentric, rather than concentric to the earth with a swelling at the antipodes. A quick look at Figs. 2.4a, b, and 2.5 shows why. "Mauritania" coincides with the western or south-western coast of Africa, at the extreme end of the habitable quarter and abutting directly the ocean toward the equator (Latini 1839: 175). Observers positioned there will be able to see the water rise in front and *above* them, rather than *below*. This cannot happen in Fig. 2.4a, where the bulging of the sphere of water is located at the antipodes and thus remains forever invisible to the dwellers of the inhabitable quarter. From the point of view of the Mauritanian observer, this model and Dante's favorite are indistinguishable.

Latini seems to have been an important reference for the Tuscan merchant who, according to Rita Librandi, might have been the interpolator of a manuscript of the vernacular *Meteorology* composed in Florence around 1310 (Librandi 1995: I, 45–54). While this work was proposed as a popularization of Thomas Aquinas's commentary, the anonymous reader appears well aware that most of the material came instead from Albertus Magnus's. Undeterred by the fame of the Dominican scholar, the interpolator did not hesitate to criticize him when the text seemed to contradict empirical observations, including his own. The merchant's travels in the "Sea of England," for instance, showed him that Albertus's doctrine of the reflux of the Ocean could not be correct.[16] Experience supported, too, the counterintuitive but widespread notion that the surface of the ocean is higher than "any high mountains on the earth." As fellow "merchants of Genoa" who "were there" (or perhaps, Brunetto Latini?) allegedly told him, from the Atlantic coast of Africa the ocean appears "visibly very high and it towers above the shore, and it scares those who see it for fear that it overruns the earth."[17] There was no need to travel or to trust Genoese merchants, though, to become convinced of the physical possibility of the phenomenon: just "take a circular vase, and fill it completely with water; you will see that its level is higher in the middle than along the rim."[18]

Latini did not explain in detail why the ocean remained self-contained, against its natural tendency to form a sphere around the center of the universe, but the anonymous interpolator and Restoro of Arezzo did. In his vernacular *La composizione del mondo* (completed in 1282) the Tuscan artist merged his bodily and empirical experience of nature with a robust theoretical framework, mainly derived from Latin translations from the Arabic that emphasized astrological science as the main tool for the understanding of nature—immediately followed, in his opinion, by drawing, engraving and painting (Restoro d'Arezzo 2007: 50).[19] Like the author of the *Questio*, Restoro conceived dry land as a half moon-like section of earth emerging from a larger sphere of water. Unlike him, he does not even discuss the possibility that the earth could just have a hump: his only question is which of the two spheres is eccentric (Restoro d'Arezzo 2007: 177). Restoro surmised that stars, just like magnets, could attract but even repel and that the northern stars are more numerous and powerful (2007: 175–76). As their power travels across the elemental spheres that compose the inferior world, by necessity it encounters water before earth, removing it from the center of the world just enough to uncover "its fourth part, so that three remained underwater and [the learned] called it the inhabitable quarter." This sounds all the more plausible, as water is lighter than earth and easier to move:

> Moreover, water is lighter than earth: and it is therefore reasonable that it be pushed back rather than the heavier earth moved upwards; therefore we have the earth uncovered, and water pushed back according to reason. And if the virtue of the sky, that must push water back so as to keep the earth uncovered, ceased and disappeared, the removed water would come back to its place and cover the whole of the earth.[20]

The "virtue" of the northern stars acts upon the earth as well, which is on the contrary attracted toward them; while this element is heavier and harder to move, their action is sufficient for the generation and continuing existence of mountains. Since those constellations are unevenly populated with stars of differing brightness and power, their attraction is uneven too leading to elevations, valleys, and plains:

> In the appropriate time, the heavens by their virtue can gather mounds of earth, heap them one upon the other [geological strata?] and attract this earth towards the sky, like the magnet by its virtue can attract iron, making as many mountains and as high as needed, like a blacksmith lays down layers of iron one upon the other in order to make the anvil he needs.[21]

The displacement of the ocean remained the sole responsible for the existence of the inhabitable quarter since such mountains did not alter sensibly the sphericity of the earth. One may wonder why those texts could not assume different shapes for the sphere of water apart from those examined in the *Questio*—say, one that bulges right before the

Mauritanian shores but is elsewhere concentric to the earth. This is in fact highly unlikely, as such variants would defeat the very reason why the doctrine of the higher elevation of the ocean existed in the first place: once admitted, the eccentricity of the sphere of water (or the arguably less common antipodal swelling) provided a simple, elegant, and physically cogent explanation to the problem of the existence of water sources on mountains.

The Subterranean World

From antiquity to about 1700, it was commonly accepted that much of the cycle of water takes place in the underground. Snow, rainfall, or the condensation of surface humidity could not account for part or most of the water found in springs and rivers so the remainder must come from subterranean sources. Most explanations rested on the porous nature of mountains or on a network of subterranean veins, channels, and caverns pervading, if not the whole of the earth, the side or the swelling that emerged from the ocean (Fig. 2.3).

Aristotle maintained that mountain springs originated from both the percolation of water from the soil, and the action of exhalations within the earth that turned rainwater into wet vapors; it was then their nature to move upward through the volume of the mountains, cooling down as they reach higher regions. As usual, the Philosopher was skimpy on details and a number of authors set out to integrate his account, or to elaborate alternatives. Drawing upon traditions that circulated already in Aristotle's time and in Roman authors, Bartholomew the Englishman believed that rivers proceeded exclusively from subterranean channels connecting the sea bottoms to water springs in the inhabitable quarter (Sarti 2019: 4). At the other end of the spectrum, Albertus Magnus posited that they originated from rainfalls that mountains absorb as if they were sponges. Water accumulated inside them in vast subterranean caverns full of exhalations, which again caused its transformation into wet vapors thus explaining the persistence of rivers even during dry seasons (Sarti 2019: 6–7).

Four centuries later, Athanasius Kircher's lavishly illustrated *Subterranean World* still described an earth pervaded by underground streams flowing from the seafloor to immense reservoirs located in the mountains' hollow bodies, which in turn fed the rivers that returned water to the sea. The inextinguishable fire burning in the center of the earth, and its smaller subsidiaries closer to the surface, provided the heat that sustained the whole cycle (Kircher 1665). Yet the idea of a central fire only took hold in the course of the sixteenth and seventeenth centuries (Vermij 1998; Ciancio 2018a, 2018b). This subterranean heat was not available to medieval meteorology in which the center of the globe was normally cold, if not the coldest place in the universe being the furthest from the Sun and shielded from its rays by the whole mass of the earth. In its absence, what power could lift water from the seabed to a mountaintop? Alternatively, how could hot exhalations form? While answers could be given, they all brought their own sets of problems. However, the pressure exerted by a sphere of water whose surface is higher than every mountain could provide the desired mechanism.

The principle could be easily grasped by anyone who understood the operation of an aqueduct or an artificial fountain. If the reservoir is higher than the spigot, water will rise into the pipes and gush out onto the surface despite being a heavy element. Once there, it resumes its normal downward motion according to its nature. Water behaves in the same way on a global scale: being eccentric to the center of gravity of the universe, the sphere of water weighs more on one side than on the other thus exerting an enormous pressure in the opposite direction. Water then tends to rush to the uncovered part of the earth through its porosities and the network of subterranean channels, which was often likened to veins and arteries in the human body. Once there, it escapes to the surface where it is no longer subject to that force and can flow *downward* toward the ocean, which along the coastlines is *lower* (namely, closer to the center of gravity of the universe) than the rest of the habitable quarter. The simplicity of the theory is compelling, and its physical principles indisputable: as long as the surface of the ocean is higher, water will rise onto even the tallest mountains. Neither invisible exhalations nor hypothetical sources of heat need to be involved. Figure 2.5 shows why the swelling or gibbosity of the sphere of water must be located at the antipodes of the dry land: as water tries to restore its concentricity, it exerts its pressure toward the opposite side passing through the center of gravity of the universe. The fluid can only move in the same direction.

Medieval sources leave no doubts that the strength of this doctrine lay in its ability to explain the existence of fountains on the emerging quarter of the earth. In fact, the phenomenon was often cited as evidence that the sphere of water must be higher than that of earth. As Brunetto Latini explains:

> The earth is all hollowed here and there, and full of veins and caverns, so that the waters that leave the sea come and go through the earth, and in and out according to where the veins carry them, just like a man's blood spreads through its veins, reaching the whole body upwards and downwards... And if it is true that the water rests upon the earth, then it must be higher than the earth. And if the sea is higher than the earth, it is little marvel that fountains spring even on the highest mountains.[22]

Likewise, Restoro describes how sea water penetrates the earth and circulates underground where it is filtered losing its salinity before it reaches the surface forming streams and rivers (Restoro 2007: 184). Having disproved what looks like Albertus Magnus's doctrine on the existence of large reservoirs inside the mountains—but without naming any authors (2007: 192)—the Aretine artist proposes what he regards as a more plausible cause: "But water can well raise to mountains in this way, namely because being water spherical, and having to cover the circumference of the earth according to reasons, in this way water will be higher than earth."[23]

The author of the *Questio* mentions this argument as the fourth in the list of the reasons brought up by his opponents, exposing it with the usual clarity: "since the sea is the source of all waters (as is made clear by the Philosopher in his *Meteorics*) if the sea were not higher than the earth, the water would not flow to the earth – for in every natural movement of water the source seems to be higher" (Alighieri 1903: 11–13).[24] Dante's strongest objections to the theory that the sphere of water is either eccentric or gibbous come down, essentially, to water being a heavy element and a fluid that cannot contain itself. As such, it will naturally flow downward until every point of its surface has the same distance from the center of the universe (Alighieri 1979: 751–55): "And if any one should deny these two axioms [water always flows downward, and it is a fluid], or either thereof, it would be impossible to convince him... because truly these axioms are founded on observation and induction" (Alighieri 1903: 17).[25]

In fact, neither Brunetto Latini, nor Restoro of Arezzo, nor the anonymous interpolator dreamt of denying them. On the contrary, Restoro contended that the sphere of water clears the inhabitable quarter only because it is pushed back by the power of the northern stars *against* its nature and that it would immediately submerge the dry land should this force cease (Restoro d'Arezzo 2007: 178). One can only speculate as to why Dante sets out to demonstrate a truism, while ignoring more sophisticated arguments that go entirely unanswered. Perhaps, his Mantuan interlocutors were ignorant of them as well. The authors of the popular encyclopedias that may have been their sources often explained that water is higher than earth; cited observations and other evidence for the reality of the phenomenon; but usually fell short of explaining how it happened. Latini did not go into any details either. The anonymous interpolator seems to think that water can actually self-contain itself without running over the dry land, as in his example of the vase filled with water, but like Restoro he also cites the influence of the planets as the reason why the sea uncovers part of the earth.[26] The least one can say is that Dante is unlikely to have read Restoro or his sources, and in general he does not seem to know or understand the finer points of the doctrine he sets out to confute.

This impression is reinforced by some of the empirical evidence he mentions in support of his thesis. The first one is that rivers and streams flow everywhere toward the sea; if water were higher than earth, then they would be running *upwards*—again, entirely against common sense and the nature of the element of water:

> For as to observation, we see throughout the earth rivers flowing down to the sea, in the south and in the north, in the east and in the west, which could not occur if the sources of the streams and the course of their channels were not higher than the surface of the sea. (Alighieri 1903: 13)[27]

However, this is not what happens. As can be seen in Figs. 2.4b and 2.5, all along the coastlines of the inhabited quarter the surface of the ocean is actually *lower* than the shore, with the distance from the center of the universe steadily increasing as one moves further away from the emerged quarter. This means that on the continent water flows *downward* according to its nature, as it should once it is no longer subject to the pressure of the sphere of water. Neither Latini nor Restoro claim or imply otherwise. The second observation is a bit puzzling too. Among the arguments used by the opponents, the *Questio* mentions the experience of those sailors who "when at sea behold mountains beneath them":

> And to prove that they say that climbing the mast they see them, whereas indeed from their decks they do not see them; which seems to be because

> the earth is really lower than, and sunken beneath, the crest of the sea. (Alighieri 1903: 11)[28]

Dante is right of course to point out that this observation does not prove that water is higher, but only that it has a spherical shape—or to nitpick, that it is convex at the observer's location (Alighieri 1979: 770). This was indeed one of the canonical proofs of the sphericity of the Earth, from Aristotle to John of Holywood's *On the Sphere*. This particularly weak piece of empirical "evidence" is nowhere to be found in Latini, Restoro, or the interpolator, though it might be present in literature I have not perused and was perhaps brought up in the Mantuan disputation. Latini and the interpolator talk instead of the observation that from the Mauritanian coast, the ocean would appear higher than the shore—information which the *Questio* neither mentions nor questions. Overall, the part of the lecture that refutes the eccentricity or bulging of the sphere of water seems to take on rather inconsequential issues, while weightier arguments that were available in coeval literature remain unaddressed. Only toward the end, the author goes back briefly to one of the most important, namely the existence of water sources on even the highest mountains:

> For the vulgar herds and those ignorant of physical arguments think that water ascends to the summit of mountains and even to places where fountains are, in the form of water; but this is surely puerile, for the waters are generated there – as the Philosopher shows in his Meteorics – the material ascending in the form of vapors. (Alighieri 1903: 55)[29]

The appeal to authority to dismiss the subterranean circulation of waters as "puerile" is Dante's only argument here, and possibly the reason why Aristotle is invoked even though much of his meteorology seems to come instead from Albertus Magnus (Cornish 2004; Sarti 2019).[30] The reference to the "vulgares" might hint at the popularity of that doctrine in didactic literature, and among readers of vernacular encyclopedias with only a shallow tint of learning. Yet this essay indicates that the higher elevation of water, especially in the form of the eccentricity of its sphere, was a clever and well-articulated theory that provided plausible answers to difficult physical problems—and deserved perhaps a more comprehensive confutation.

Conclusion

The scholars who have written on the eccentricity of the sphere of water have usually followed the *Questio* in its harsh judgment of that model. At a superficial glance, this construct can easily appear ridden with logical and empirical fallacies, and unworthy of serious consideration; yet a more careful analysis shows that there are good reasons why it was widely entertained for centuries. I would argue that its adoption and persisting popularity demonstrate not faulty reasoning, but intellectual sophistication. A number of medieval authors and their readers were willing to accept the physical hypotheses that best explained the observed world, even when they appeared highly counterintuitive and seemed to go against the perceived "common sense." The motion of the Earth several centuries later, or Einstein's relativity one hundred years ago, was at least as disorienting in their times if not more.

While this chapter looks at the *Questio* as a source to better understand and contextualize the doctrine it attacks, rather than the other way around, it also sheds some light on the famous lecture and its author. In particular, it lends support to the majority view that the *Questio* could have actually been written in the early fourteenth century rather than many decades later. It is difficult indeed to make sense of Latini's or Restoro's discussions of the problem unless their reasonings are underpinned already by the eccentricity of the sphere of water. Much has been written, too, on the sharp contrast between the imaginative and theological cosmology of the *Commedia* and the dry physical and geometrical argumentations of the *Questio*. In some cases, this has been regarded as yet another reason to doubt its traditional attribution. Conversely, Francesco Mazzoni argued that the *Questio* indicates an increasing concern for physical and scientific problems in Dante's late years, and even a dramatic shift in the poet's cosmological conceptions (Alighieri 1979: 727–32).

It seems unlikely, though, that Dante did not develop a keen interest in the meteorological issues debated in the *Questio* until he was in his fifties. In her works on Dante's meteorology, Alison Cornish pointed out that the discipline was pervasive in late medieval culture as attested by hundreds of Latin manuscripts and vernacular translations. Meteorological topics were often addressed even in books focusing primarily on other subjects, as well as being ubiquitous in commentaries and didactic encyclopedias: Latini's *Tresor* and Restoro's *Composizione del mondo* belonged to the latter category (2004). Since the field of meteorology dealt with everyday phenomena that were easily experienced, it was considered especially suitable as an introduction to natural philosophy for a vernacular or philosophically illiterate public. It is hardly a coincidence that the *Meteorology* was the first of Aristotle's works on natural philosophy to be translated into French and Italian vernaculars (Ducos 2001; Monfrin 1996; Librandi 1995; Dal Prete 2016). My own recent work has only reinforced Cornish's assessment, showing that the main tenets of the meteorological tradition were common knowledge among late medieval and early Renaissance craftsmen, artists, and builders.[31] There seems to be little doubt that Dante possessed a solid meteorological culture decades before the composition of the *Questio*, and that this document can hardly be seen as a demonstration of a profound change of interests and worldview in the years immediately preceding his death. If anything, the evident gaps in his knowledge of a theory especially popular in vernacular contexts may confirm that Dante was "neither an avid reader nor a benevolent judge of the translations of his time."[32]

It is perhaps more sensible to conclude with John Freccero (1961), that the poet-theologian and the natural philosopher simply described the same phenomena in different terms and with different aims, which in medieval culture were by no means mutually exclusive. The emergence of dry land from the waters had both natural and theological explanations, which in *Inferno* are also couched in the language of poetic fiction and imagination. It would have been inappropriate to do so in a philosophical discussion where, as Albertus Magnus argued, "the truth is explained as it is" and according to natural reasons (Albert the Great 2010: 87).

Notes

1. I am aware of course of the anachronism of terms like "science" or "scientific" in the sense they are normally used today. In this chapter, they are shorts for "natural philosophy" or "pertaining to natural philosophy". "Globe" is the body composed by the spheres of earth and water, same for "Earth" capitalized; "earth" is the element.
2. In line with mainstream scholarship, in this chapter I assume that Dante was indeed the author of the *Questio*. While I touch upon issues of attribution, I do not discuss them in any depth.

3. Alighieri (1979: 744): "Questio igitur fuit de situ et figura sive forma duorum elementorum, atque videlicet et terre... Et restricta fuit questio ad hoc, tanquam ad principium investigandae veritatis, ut quereretur utrum aqua in spera sua, hoc est in sua naturali circumferential, in aliqua parte esset altiora terra que emergit ab aquis et quam comuniter quartam partem habitabilem appellamus." Whenever possible, I quote from established English versions rather than provide my own translations. When not otherwise noted, translations are mine. For the *Questio*, I have normally worked on Mazzoni's edition (1979), and the *editio princeps* (Alighieri 1508).
4. Bruno Nardi (1959: 434) argued that: (a) medieval Christian philosophers saw a contradiction between the "Aristotelian doctrine that places water and the other elements concentric to the earth, and the biblical statement that God himself gathered all of the water in a single place"; (b) the problem was in fact inexistent for the Greeks and the Arabs. I believe Duhem (1958) provided ample evidence that (b) is not accurate. As for (a), I would argue that the contradiction was not between Aristotle' theory of natural place and the Bible, but between Aristotle's theory of natural place and the obvious observation that dry land does exist—whatever or Whoever the reason.
5. See also Duhem (1958: 9.269–70). For a modern edition: William of Conches (1974). William also dealt with the creation of the elements in the philosophical dialogue *Dragmaticon Philosophiae*, available in English translation (William of Conches 1997: 13–29). See also Obrist and Caiazzo (2011).
6. At a later time, the main source for this doctrine seems to have been Alexander of Aphrodisias's commentary on Aristole's *Meteorology*, translated by William of Moerbeke in 1260 (Duhem 1958: 9.239; Alexander of Aphrodisias 1968: vii). Alexander of Aphrodisias's commentary was an integral part of university teaching in Paris by 1275. It is not clear to me why Duhem (1958: 9.121–22) calls this doctrine "puérile."
7. Among the ample literature on Bernard Silvestris and his *Cosmographia*, see Stock (2015), Kauntze (2009, 2014), Dronke (1980). Among the editions and translations of his work, see Michel Lemoine's Bernard Silvestris (1998), Peter Dronke's

Bernardus Silvestris (1978), Winthrop Whetherbee's Bernard Silvestris (2015).

8. Albertus Magnus, *Liber Methauroru[m]*, Liber 2, Tractatus 3, Capitulum 1.
9. A "vernacular meteorology" appears in one of the two remaining lists of Leonardo's books (Librandi 1995: 2.268). While this work proposed itself as a translation of Thomas Aquinas' own commentary, the bulk of the material comes from Albertus Magnus's. Leonardo's most explicit reference to this theory is in London, British Library, MS Arundel 263, fol. 155v. See Marcolongo (1926: 1.176–77) (I could not consult the 1998 edition by Carlo Pedretti and Carlo Vecce). Leonardo touched again upon this concept in a passage dealing with astronomy: Paris, Bibliothèque de l'Institut de France, MS 2177 (Manuscript F), fol. 69v, on which see Marinoni (1988: 115).
10. Alberti's was a common image among the supporters of this theory.
11. Rutkin's work on medieval astrology and natural philosophy is a must read. Still, scholars have largely focused on the formation and properties of minerals, metals and fossils. The role of astrological forces in the global geological aspect and evolution of the Earth remains comparatively understudied, but see Dal Prete (2022: 47–51).
12. A rough sketch of Buridan's theory of geological change appeared first in his commentary *On the Heavens*, but a fuller account was given later where it belonged most: in the *Questions on the First Three Books of the Meteorology* (c. 1352).
13. Buridan (1996: 417): "Cum ergo illae partes interiores terrae elevantur modo praedicto ad superficiem terrae, illae quae sunt teneres et divisibiles per ventos et pluvias et fluvios, iterum moventur ad profundum maris; aliae se autem magis durae et lapidosae non possunt sic dividi et fluere, et ideo manent et continue per longissima tempora elevantur per totalem terrae elevationem; et ita possunt fieri montes altissimi. Et, si nulli essent modo montes, adhuc per istum modum fierent in futuro; nec apparent alii modi per quos possent generari et manere tales montes." On Buridan's doctrine, see also Duhem (1958: 9.196–202, 295–96), Dal Prete (2022: 57–60).
14. Buridan (1996: 414): "Et aliqui laboraverunt in mari ad permeandum mare in aliis quartis, et numquam poterunt pervenire ad

aliquam terram habitabilem; et ideo dicitur quod Hercules in finibus huius quartae infixit columnas, in signum quod ultra eas non erat terra habitabilis nec mare permeabile". Also in Buridan (1942: 158). See also Duhem (1958: 9.197), Moody (1941), Boffito (1905).

15. Latini (1839: 175): "Che 'l mare v'è assai più alto che la terra, e tiensi sì in fra sé, che non cade né corre sopra la terra."
16. Librandi (1995: 1.326). Other interventions hint at business travels to other areas of economic interest for Florentine companies, like Naples and Greece.
17. Librandi (1995: 1.324): "Ancora avemo saputo da mercatanti genovesi degni di fe', che vi sono stati, che nel paese di Libia Mauritana, in sulla rive del Mare Occeano, dicono che il mare visibilmente è altisimo e soperchia la riva, a e ffa paura a cchi'l vede che no trabocchi sopra la terra: e ser Brunetto ne fa menzione del suo *Tesoro*, ove divisa i confini e' paesi d'Africa."
18. Librandi (1995: 1.324): "Ancora per exempro visibile tolli uno vaso tondo ed empilo a raso d'aqua quanto tiene, e vedrai che'l colmo dell'acqua del mezzo del vaso è più alto che quella delli orli del vaso."
19. Recent scholarship has refuted the traditional view that the author of the *Composizione* would have been a monk. All we know about him (admittedly, not much) points to the painter, goldsmith, and possibly illuminator and bell founder with the same name active in Arezzo in the same decades. See Donato (1996), Altieri Biagi (1990: 13). On Restoro's sources, see Austin (1913).
20. Restoro d'Arezzo (2007: 178): "E anco l'acqua è più leve de la terra: è rascione ch'ella debbia anti èssare cessata che la terra ch'è più grave, debia èssare levata sù; e avemo la terra scoperta, e l'acqua de rascione cessata via. E se la virtute del cielo, che dea tenere l'acqua cessata per mantenere la terra scoperta, se cessasse e andasse via, l'acqua cessata tornarea en suo loco e coprirea tutta la terra."
21. Restoro d'Arezzo (2007: 127): "Adonqua è mestieri che'l cielo colla sua virtude, la quale ha ' adoparare elli monti li quali li so' mestieri come la'ncugine al fabbro, ch'elli colla virtude sua quando li fosse mestieri possa aunare la terra ensieme e amolare l'una sopra l'altra e trare envorso sé la terra, come la calamita per virtude tra' a se lo ferro e fare li monti tanti e si alti quanto li so' mestieri e anco come lo fabro, che auna l'uno ferro sopra l'altro e farne

la'ncugine la quale li è mestieri." While the reference to magnet-like attractions had been commonplace for centuries, the image of the blacksmith and of his anvil seems to reflect Restoro's artisanal (and much vaunted) background. Clearly, such mountains do not alter sensibly the sphericity of the earth and the displacement of the ocean is solely responsible for the existence of the inhabitable quarter.

22. Latini (1878: 317–18): "La terra è tutta cava dentro di luogo in luogo ed è piena di vene e di caverne e però le acque che di mare escono vanno e vegnono per la terra, e surgono dentro e di fuori secondo che le vene le menano qua e là; così come il sangue dell'uomo si sparge per le sue vene, sì che cerca tutto il corpo da monte e da valle... E se ciò è vero che l'acqua seggia in su la terra, dunque è ella più alta che la terra. E se il mare è più alto che la terra, dunque non è meraviglia delle fontane che escono su nell'alte montagne."
23. Restoro d'Arezzo (2007: 193): "Ma pò salire bene l'acqua e lo monte secondo questa via per rascione che, cum ciò sia cosa che l'acqua sia sperica e per rascione debia coprire tutta la terra entrono entorno, secondo questa via sarà l'acqua più alta de la terra."
24. Alighieri (1979: 746): "Et cum mare sit principio omnium aquarum ut patet secundum Philosophus in Metauris suis, si mare non esset altius quam terra, non moveretur aqua ad ipsam terram, cum in omni motu naturali aque principium oporteat esse altius."
25. Alighieri (1979: 750): "Et si quis hec duo principia vel alterum ipsorum negaret, ad ipsum non esset determinatio... Sunt etenim haec principia inventa sensu et inductione."
26. Librandi (1995: 1.325): "È di nicissità e per lo corso celestro e forza di pianeti che parte della terra sia scoperta."
27. Alighieri (1979: 748): "Ad sensum enim videmus per totam terram flumina descendere ad mare, tam meridionale quam septentrionale, tam orientale qual occidentale; quod non esset, si principia fluminum et tractus alveorum non esset altiora ipsa superficie maria."
28. Alighieri (1979: 746): "Secunda sive minor per experiential nautarum, qui vident, in mari existentes, montes sub se, et proban dicendo quod ascendendo malum vident eos, in navi vero non vident; quod videtur accidere propter hoc, quod terra valde inferior sit et depressa a dorso maria." See also Alighieri (1979: 754).

29. Alighieri (1979: 77): “Credunt enim vulgares et physicorum documentorum ignari quod aqua ascendat ad cacuminal montium et etiam ad locum fontium in forma aque; sed istud est valde puerile, nam aque generantur ibi, ut per Phylosophym patet in Metauris suis, ascendente materia in forma vaporis.”
30. I am grateful to Edward Allnutt for our stimulating conversations on the subject.
31. Dal Prete (2022: 65–92). I can only agree with Cornish’s statement that “it is something of a false dichotomy, therefore, to separate learning from vernacularity” (2004: 65). See also Morgan Ng (2021), on the use of Aristotelian meteorology in the construction of earthquake-resistant buildings in Renaissance Italy.
32. Cornish (2004: 67). Citation from Concetto Marchesi.

References

Albert the Great. 2010. *On the Causes of the Properties of the Elements (Liber de Causis Proprietatum Elementorum)*. Trans. Irven Michael Resnick. Milwaukee: Marquette University Press.

Alberti, Leon Battista. 1988. *On the Art of Building in Ten Books*. Trans. Joseph Rykwert, Neil Leach, and Robert Tavernor. Cambridge, MA: MIT Press.

Alexander of Aphrodisias. 1968. *Alexandre d’Aphrodisias commentaires sur les Météores d’Aristote*, ed. A. J. Smet. Louvain: de Wulf-Mansion Centrum.

Alighieri, Dante. 1508. *Questio florulenta ac perutilis de duobus elementis aquae & terrae tractãs nuper reperta que olim Mantuae auspicata*, ed. Giovanni Benedetto Moncetti. Venice: per M. de Monteferrato.

Alighieri, Dante. 1903. *A Translation of the Quaestio de Aqua Et Terra: With a Discussion of Its Authenticity*. Trans. Alain Campbell White. Boston: Ginn & Company.

Alighieri, Dante. 1979. *Questio de aqua et terra*, ed. Francesco Mazzoni. In Dante Alighieri, *Opere minori*, t. 2, 693–737. Naples: Ricciardi.

Altieri Biagi, Maria Luisa. 1990. Nuclei concettuali e strutture sintattiche nella *Composizione del mondo* di Restoro d’Arezzo. In *L’avventura della mente: studi sulla lingua scientifica*, 11–33. Naples: Morano.

Austin, Herbert Douglas. 1913. *Accredited Citations in Ristoro d’Arezzo’s Composizione Del Mondo: A Study of Sources*. Turin: Guido Momo.

Bernard Silvestris. 1998. *Cosmographie*. Transl. Michel Lemoine. Paris: Cerf.

Bernard Silvestris. 2015. *Poetic Works*. Transl. Winthrop Wetherbee. Cambridge, MA: Harvard University Press.

Bernardus Silvestris. 1978. *Cosmographia*, ed. Peter Dronke. Leiden-Boston: Brill.

Boffito, Giuseppe. 1902. *Intorno alla "Quaestio de aqua et terra" attribuita a Dante*. Turin: Carlo Clausen.

Boffito, Giuseppe. 1905. *L'eresia degli antipodi*. Florence: Ist. alla Querce.

Buridan, Jean. 1942. *Iohannis Buridani Quaestiones Super Libris Quattuor de Caelo et Mundo*, ed. Ernest A. Moody. Cambridge, MA: The Mediaeval Academy of America.

Buridan, Jean. 1996. *Ioannis Buridani Expositio et quæstiones in Aristotelis De cælo*, ed. Benoît Patar. Louvain-La-Neuve: Editions Peeters.

Ciancio, Luca. 2018a. An Amphitheatre Built on Toothpicks: Galileo, Nardi and the Hypothesis of Central Fire. *Galilaeana* 15: 83–113.

Ciancio, Luca. 2018b. 'Immoderatus Fervor Ad Intra Coërcendus': Reactions to Athanasius Kircher's Central Fire in Jesuit Science and Imagination. *Nuncius* 33 (3): 464–504.

Cornish, Alison. 2004. The Vulgarization of Science: Dante's Meteorology in Context. In *Science and Literature in Italian Culture: From Dante to Calvino*, ed. Pierpaolo Antonello and Simon Gilson, 53–71. Cambridge, UK: Legenda.

Dal Prete, Ivano. 2016. Vernacular Meteorology and the Antiquity of the World in Medieval and Renaissance Italy. In *Vernacular Aristotelianism in Italy from the Fourteenth to the Seventeenth Century*, ed. Luca Bianchi, Jill Kraye, and Simon A. Gilson, 139–59. London: The Warburg Institute.

Dal Prete, Ivano. 2022. *On the Edge of Eternity: The Antiquity of the Earth in Medieval and Early Modern Europe*. Oxford and New York: Oxford University Press.

Dales, Richard C. 1965. Anonymi De Elementis: From a Twelfth-Century Collection of Scientific Works in British Museum MS Cotton Galba E. IV. *Isis* 56 (2): 174–189.

Dales, Richard C. 1990. *Medieval Discussions of the Eternity of the World*. Leiden and New York: Brill.

Donato, Maria Monica. 1996. Un 'savio depentore' fra 'scienza de le stelle' e 'sutilita' dell'antico. Restoro d'Arezzo, le arti e il sarcofago romano di cortona. *Annali della Scuola Normale superiore di Pisa. Classe di Lettere e Filosofia*, IV/1–2: 52–78.

Dronke, Peter. 1980. Bernard Silvestris, Natura, and Personification. *Journal of the Warburg and Courtauld Institutes* 43: 16–31.

Ducos, Joëlle. 1998. *La météorologie en français au Moyen Age (XIIIe–XIVe siècles)*. Paris and Genève: Honoré Champion.

Ducos, Joëlle. 2001. L'oeuvre de Mahieu le Vilain: Traduction et commentaire des Météorologiques. In *Les traducteurs au travail. Leurs manuscrits et leurs méthodes*, ed. Jacqueline Hamesse, 285–309. Turnhout: Brepols Publishers.

Duhem, Pierre. 1958. *Le système du monde: histoire des doctrines cosmologiques de Platon à Copernic*, 10 vols. Paris: A. Hermann.

Fioravanti, Gianfranco. 2019. Alberto di Sassonia, Biagio Pelacani e la *Questio de aqua et terra*. *Studi Danteschi* 89: 81–97.

Freccero, John. 1961. Satan's Fall and the *Quaestio de Aqua et Terra*. *Italica* 38 (2): 99–115.

Kauntze, Mark. 2009. The Creation Grove in the 'Cosmographia' of Bernard Silvestris. *Medium Ævum* 78 (1): 16–34.

Kauntze, Mark. 2014. *Authority and Imitation: A Study of the Cosmographia of Bernard Silvestris*. Leiden-Boston: Brill.

Kaye, Joel. 2014. *A History of Balance, 1250–1375: The Emergence of a New Model of Equilibrium and Its Impact on Thought*. Cambridge, UK: Cambridge University Press.

Kircher, Athanasius. 1665. *Athanasii Kircheri... Mundus Subterraneus: In XII Libros Digestus*, 2 vols. Amstelodami: apud J. Janssonium & E. Weyerstraten.

Latini, Brunetto. 1839. *Il Tesoro di Brunetto Latini volgarizzato da Bono Giamboni, nuovamento pubblicato secondo l'edizione del MDXXXIII*. Venice: Co' tipi del Gondoliere.

Latini, Brunetto. 1878. *Il tesoro, volgarizzato da B. Giamboni, emendato ed. illustr. da L. Gaiter*, ed. P. Chabaille and Luigi Gaiter. Bologna: Gaetano Romagnoli.

Librandi, Rita, ed. 1995. *La Metaura d'Aristotile: volgarizzamento fiorentino anonimo del XIV secolo*, 2 vols. Naples: Liguori.

Marcolongo, Roberto, ed. 1926. *I manoscritti e i disegni di Leonardo Da Vinci. Il Codice Arundel 263*, 5 vols. Rome: Danesi.

Marinoni, Augusto, ed. 1988. *I manoscritti dell'Institut de France*, Manoscritto F. Florence: Giunti Barbèra.

Martin, Craig. 2011. *Renaissance Meteorology*. Baltimore: Johns Hopkins University Press.

Monfrin, Jacques. 1996. Jean de Brienne, comte d'Eu, et la traduction des Météorologiques d'Aristote par Mahieu le Vilain (vers 1290). *Comptes Rendus Des Séances De L'académie Des Inscriptions Et Belles-Lettres* 140 (1): 27–36.

Moody, Ernest A. 1941. John Buridan on the Habitability of the Earth. *Speculum* 16 (4): 423–424.

Nardi, Bruno. 1959. *La caduta di Lucifero: e l'autenticità della "Quaestio de aqua et terra."* Turin: Società editrice internazionale.

Ng, Morgan. 2021. Terremoti artificiali. La sismologia aristotelica nella guerra sotterranea del Rinascimento. In *Material World. The Intersection of Art, Science, and Nature in Ancient Literature and its Renaissance Reception*, 133–62. Leiden: Brill.

Nothaft, C. Philipp E. 2014a. Origen, Climate Change, and the Erosion of Mountains in Giles of Lessines's Discussion of the Eternity of the World (c. 1260). *The Mediaeval Journal* 4 (1): 43–69.
Nothaft, C. Philipp E. 2014b. Climate, Astrology and the Age of the World in Thirteenth-Century Thought: Giles of Lessines and Roger Bacon on the Precession of the Solar Apogee. *Journal of the Warburg and Courtauld Institutes* 77: 35–60.
Obrist, Barbara, and Irene Caiazzo, eds. 2011. *Guillaume de Conches: philosophie et science au XIIe siècle*. Florence: SISMEL Edizioni del Galluzzo.
Pellegrini, Paolo. 2021. La *Questio de aqua et terra* di Dante: storia e storiografia. *Rivista Di Filosofia Neoscolastica* 113 (1): 183–206.
Restoro d'Arezzo. 2007. *La composizione del mondo*, ed. Alberto Morino. Lavis: La Finestra Editrice.
Rutkin, Darrel H. 2019. *Sapientia Astrologica: Astrology, Magic and Natural Knowledge, ca. 1250–1800: I. Medieval Structures (1250–1500): Conceptual, Institutional, Socio-Political, Theologico-Religious and Cultural*. Cham: Springer International Publishing.
Sarti, Luna. 2019. Dante and the Water Cycle in the *Commedia*: God, Physics, and Flowing Waters. *Dante Studies* 137 (1): 1–22.
Stock, Brian. 2015. *Myth and Science in the Twelfth Century: A Study of Bernard Silvestris*. Princeton, NJ: Princeton University Press.
Tanner, Norman P. ed. 1990. *Decrees of the Ecumenical Councils*, 2 vols. London-Washington, DC: Sheed & Ward-Georgetown University Press.
Vermij, Rienk. 1998. Subterranean Fire. Changing Theories of the Earth During the Renaissance. *Early Science and Medicine* 3/4: 323–47.
William of Conches. 1531. *Philosophicarum et astronomicarum institutionum, Guilielmi Hirsaugiensis olim abbatis, libri tres*. Basel: Heinrich Petrus.
William of Conches. 1974. *Philosophia mundi: Ausgabe des. 1. Buchs von Wilhelm von Conches' "Philosophia" mit Anhang, Übersetzung und Anmerkungen*, ed. Grealgor Maurach. Pretoria: University of South Africa.
William of Conches. 1997. *A Dialogue on Natural Philosophy (Dragmaticon Philosophiae)*, ed. Italo Ronca and Matthew Curr. Notre Dame: University of Notre Dame Press.

CHAPTER 3

Passion and Passionless Love in Dante: From *Epistola* 3 to the *Commedia*

Joseph Romano

Between 1303 and 1306, shortly before he began the *Commedia*, Dante wrote *Epistola* 3 and *Io sono stato con amore insieme*, a letter and sonnet addressed to fellow poet Cino da Pistoia (Baglio 2016: 80–89; Toynbee 1966). In the letter Dante responds to the question Cino had previously set him: "utrum de passione in passionem possit anima transformari,"

At every turn, this chapter has benefitted from discussions with Rohini Ralby, whose expertise in contemplative practice and the theology of the unitive love of God informs throughout this chapter's engagement with desire, will, and passionless love in Dante, especially in the sections on the canzoni, Statius, St. John, and the final image. I extend her my deepest thanks. I also sincerely thank David Soud, Teodolinda Barolini, Eleanor Johnson, Christopher Cullen, Franco Masciandaro, and Nicola Masciandaro for their invaluable feedback and suggestions. I would also like to thank Christopher Baswell, Emily Runde, Russell Friedman, and Tuomas Vaura for their generous help with the manuscripts and transcriptions of the work of Peter of Trabibus discussed below. Full bibliographical references of the editions and translations used are in References.

M. Pace (ed.), *Dante and the Sciences of the Human*, Palgrave Studies in Medieval and Early Modern Medicine,
https://doi.org/10.1007/978-3-031-69253-6_3

whether the soul can be transformed from one passion to another (*Ep.* 3.2). In his reply, Dante makes an unusually overt pronouncement on the "sedes amoris" or seat of love: "potentia concupiscibilis, que sedes amoris est, sit potentia sensitive" (*Ep.* 3.2). Love arises in the soul, Dante asserts, in the "concupiscible power," the faculty of appetition aroused by and directed toward objects of sense perceived as pleasurable, suitable goods. From this premise that love is seated in the sensitive soul and that all sense-based faculties and phenomena are corruptible, Dante draws the conclusion that love itself is corruptible and mutable, governed by the Aristotelian principle *corruptio unius generatio sit alterius*, the passing-away of one thing is the coming-to-be of another (*Ep.* 3.4).[1] Therefore, Dante responds in the affirmative to Cino's question: given that love is a passion of the sensitive soul, yes, the soul may be transformed from one passion of love to another. One love may corrupt and another may generate in its place (*Ep.* 3.5).

It is "remarkable," write Foster and Boyde, that Dante considers love a corruptible passion in *Epistola* 3 because already in the *Vita nuova* he had written of "an entirely sublimated 'heavenly' love," and in canzoni, especially those included in the *Convivio*, he "celebrated a love that had its seat in the mind or intellect" (1967: 2.323).[2] Following Dante's own dictum, that by speaking of "the place in which love operates, we understand the kind of love it is," this chapter begins with a commentary on *Epistola* 3's Aristotelian phenomenology of love as a material and mutable passion seated in the sensitive soul. Demonstrating Dante's philosophical originality, especially as compared to his scholastic contemporaries, and extending the work of Teodolinda Barolini, this chapter then shows how the Third Epistle is an inflection point in the shifting definitions of love across Dante's career (Barolini 2022: 244–45).[3] While *Epistola* 3 places love in the soul's sensitive powers, certain canzoni, the *Convivio*, and the *Commedia* locate love in the intellective powers of the soul. Here, I map Dante's evolving treatments of love along two axes: one axis denotes whether love is a material or immaterial phenomenon, and the other indicates whether love is primarily a function of intellect or desire. In *Epistola*

J. Romano (✉)
Department of English and Comparative Literature, Columbia University, New York, NY, USA
e-mail: jr3850@columbia.edu

3, love is material and non-intellective, a mutable passion of sensual desire outside the regulation of reason and will. In the canzone *Doglia mi reca*, love is also a material passion, but now it is a desire virtuously regulated by reason. *Amor che nella mente mi ragiona*, and to a greater extent, its *Convivio* commentary, delineates an immaterial dimension of love seated in the faculty of knowing rather than of appetition. In contrast, in the *Commedia* love is neither a passion of the sensitive appetite, nor a passion ordered to reason, nor an act of intellect. Rather, the *Commedia* evinces a love that is beyond passion, yet still a kind of desire: the love of God, which from the human side of things, is expressed as a spiritual affect of the will, and seated in the incorruptible dimension of the soul.

Critical accounts of intellectual love in Dante tend to have an ethical, metaphysical, or theological emphasis: seeing love as the ethical regulation of passion by reason (Barolini 2006: 47–69, 70–101; Borzi 1985; Gentili 2005; Malato 2005; Porro 2014; Sasso 2011); a metaphysical act of intellection or desire for knowledge (Ardizzone 2011; Corti 1983; Falzone 2010a; Fenzi 1999; Gagliardi 1997, 2004); or the sanctifying *caritas* Beatrice inspires in the poet (Ariani 2009; Gubbini 2011, 2017; Mocan 2018; Moevs 2005; Singleton 1977; Zambon 2022). Recently, there has also been an emphasis on the physiological and natural philosophical accounts of love in Dante (Boyde 1993; Rea 2018; Tonelli 2015). What follows demonstrates how in the *Commedia* love is not a passion seated in the sensitive appetite, as it is in *Epistola* 3, nor is love intellectual because it is seated in the intellect, as it is in the *Convivio*. Rather, the final sections of this chapter discuss four canti—the scene of Statius' liberation in *Purgatorio* 21, Beatrice's discourse on the soul in *Paradiso* 7, St. John's examination of the pilgrim on the nature of love in *Paradiso* 26, and the final image of the poem—in order to show that Dante's mature notion of intellectual love is that it pertains less to cognition than to volition. Intellectual love unites the human will to God.

The fundamental distinction between the passion of love and the passionless love proper to the will hinges on Dante's ontology of the soul. Following the division in the medieval Aristotelian tradition between the vegetative, sensitive, and rational souls, Dante considers the human soul as a tri-partite unity, "che vive e sente e sé in sé rigira," that moves, feels, and is self-aware—or more literally, turns upon itself (*Purg*. 25.75).[4] In other words, the human soul has "vegetative" or vital and nutritive powers, "sensitive" powers of movement and sensation, and "rational" powers of intellection and volition. Dante identifies the essence of the human being with its most noble part, the rational soul, the substantial form that enlivens the body, and yet is free from causal determination by the body (*Conv.* 2.7.3; *Purg.* 25.49–51). Because it is transmaterial—enmattered yet itself immaterial—the rational soul is incorruptible and remains after death (*Conv.* 2.8.7–16). The body and its organs of sensation, however, because they are material and corruptible, do pass away.

As for Dante, for Thomas Aquinas, the transmateriality of the rational soul—its immanence to and transcendence of matter—accounts for the equivocation between love as a passion of the sensitive appetite and as an affect of the will. In a section of the *Summa contra gentiles* that investigates whether after death the soul still experiences love, Aquinas warns of conflating these two senses of love:

> As for the other operations of the soul, such as loving, rejoicing, and the like, one must beware of equivocation. For sometimes such operations are taken inasmuch as they are passions of the soul, and in this sense they are acts of the sensible appetite appertaining to the concupiscible and irascible powers, entailing some bodily change. And thus they cannot remain in the soul after death, as Aristotle proves in the *De anima*. Sometimes, however, such operations are taken for a simple act of the will, in the absence of all passion... Now, since the will is a power employing no organ, as neither does the intellect, it is plain that these things of which we are speaking remain in the separated soul, so far as they are acts of the will (*SCG* 2.81.15).[5]

The difference between love as a passion and love without passion rests on the distinction between the modes of desire proper to the sense-perceptive and intellective faculties of the soul: the sensitive appetite and the rational appetite or will. Passions are acts of the sensitive appetite; more narrowly defined, they are the emotional states of the soul that arise upon the apprehension of sensible objects. Passions are "in" the soul insofar as they are modifications or "accidents" of the soul's faculties of desire/appetition. Yet all passions, including the passion of love, are also always embodied phenomena; bodily effects always accompany passion ("cum aliqua permutatione corporali," *SCG* 2.81.15). Analogous to, yet ontologically different from the passions are the "intellectual affections" (*intellectiva affectione*), which, as "simple acts of the will" (*simplici actu voluntatis*), are purely "spiritual" without bodily alteration (Lombardo 2011: 76–77). Passionless love is seated therefore in the incorruptible rational soul, not in sensation or the body.[6]

As for Aquinas, for Dante, the nature of the soul and the nature of love are intertwined. Whether love is a passion of sensitive desire or an affect of the will depends on whether the seat of love is in the material and corruptible or immaterial and incorruptible dimensions of the soul. One of the major throughlines of this chapter is to show how, most visibly in *Epistola* 3 and *Paradiso* 7, Dante marks the ontological threshold between sensitive and intellective love by way of Aristotle's account of generation and corruption, how something comes to be or passes away. In *Epistola* 3, Dante is rigorously Aristotelian and surprisingly materialist: he treats love as a passion of the sensitive soul, and he remains philosophically consistent by treating love as subject to corruption and generation. In *Paradiso* 7, Dante again applies Aristotle's principle of generation and corruption, only now to make the opposite point. Because the rational soul is created "sanza mezzo," in an unmediated divine act, the soul is safeguarded from corruption ("da corruzion sicure," *Par.* 7.129). The direct creation of the rational soul makes it incorruptible and, therefore, capable of a love that is beyond passion (*Purg.* 25.71–72; *Par.* 7.143–44).

For Dante, love that is a passion is outwardly directed to objects of sense and subject to generation and corruption, while love that is without passion is free from generation and corruption, and informs an interior act of will immediately directed toward, and ultimately in union with God, *sanza mezzo*. That Dante uses generation and corruption to mark the difference between passion and passionless love shows how he uses Aristotelian natural philosophy to articulate the experience of that which exceeds both nature and philosophy: the unitive love of God.

Passion and Love in *Epistola* 3: A Sophisticated Aristotelianism

Compared to his pre-exile work, in *Epistola* 3 Dante uses a more sophisticated Aristotelian idiom to engage questions of passion, perception, being, and change. The *Vita nuova*'s definition of love as "un accidente in sustanzia," an accident in a substance, serves as a benchmark for comparison (*VN* 25.1–2).

Passio or passion, as Erich Auerbach puts it in his historical study of the word, is derived from the verb *patior*, "to undergo" or "to suffer" (Elsky 2001: 290). In its technical Aristotelian sense *passio* refers to "everything that is passively taken in... sense impression and perception, sensation and experience" (290). And crucially for what will be explored in *Epistola* 3, *passio* also signifies "change" (290). In short, passion names the "'act' of being acted upon" (Lombardo 2011: 34).

In Aristotle, passions are a subset of accidents. Accident is a capacious term for any modification—in quantity or quality—of a substance, that is, something whose form irreducibly persists on its own (per se *subsistans*). If a form has being *simpliciter*, then an accident is what changes that form to be in a certain way, "being such" (*esse tale*). Someone who grows taller remains the same person, but the person's height (an accident) cannot exist apart from the person (a substance). Passions are accidents because passions are relational modifications; they are the qualitative changes a substance undergoes as a function of being acted upon by something else. In this broad sense, "being heated" or "being cooled" are passions just as much as the emotions of desire, fear, delight, or sorrow.[7] The *Vita nuova*'s definition of love as "un accidente in sustanzia" follows this Aristotelian distinction between substance and accident. For the Dante of the *libello*, love is an accident because "non è per sé come sustanzia"; love is not something that subsists in itself, but rather is an extrinsic quality or modification of a substance (*VN* 25.1–2). Since passion is a kind of accident, and love a kind of passion—as the discourses of poetry, natural philosophy, and medicine informing Dante attest[8]—then accident and passion are interchangeable terms in the *Vita nuova's* definition of love. Both the *Vita nuova* and *Epistola* 3 consider love to be an accident/passion; the difference is that the *libello* defines love in a static sense, as a state of the soul, whereas *Epistola* 3 advances on the *Vita nuova* by considering love in a dynamic sense, of one state or passion of love changing into another.

The philosophical tenor and rigor of *Epistola* 3 have not gone unnoticed. All critics recognize that Dante's letter to Cino resembles a scholastic *quaestio*, with Dante in the role of *magister* offering the *determinatio* or conclusion to the question "utrum de passione in passionem possit anima transformari" (*Ep.* 3.2). Corrado Calenda calls the letter's approach to love "puro razionalismo scientista," pure scientific rationalism (1995: 120). For Elisabetta Graziosi, *Epistola* 3 serves as a public demonstration of Dante's poetic and philosophical acumen: given his status as an exile, Dante writes his response to Cino with the aim of securing cultural cache in the university community of Bologna, where Cino both studied and practiced law (1997: 78). Teodolinda Barolini has gone furthest in drawing out the philosophical implications of the Third Epistle and its citation of the principle of change from Aristotle's *De generatione et corruptione*, giving it a new place of significance in Dante's *oeuvre*. She shows how Dante engages the idea of generation and corruption not only in *Epistola* 3, but from the beginning to the end of his career, from *Voi che'ntendendo* to *Paradiso* 26 (2022: 162–79, 246–65, 266–84).

Nonetheless, what has not yet been sufficiently addressed is how Dante condenses an entire web of scholastic discourse in his argument about the seat of love and whether it can receive a new passion. The below commentary on the main philosophical principles underpinning *Epistola* 3 will demonstrate its philosophical complexity, and a comparison to Dante's Florentine contemporary Peter of Trabibus will show the originality of Dante's application of generation and corruption to the passion of love. More than as merely a competent Aristotelian, in *Epistola* 3 Dante emerges as a philosopher of desire in his own right. He innovates on Aristotle and his scholastic mediators to address a question definitive for the trajectory of his own life and poetry: whether one love can take the place of another, or as will be shown, whether another love can replace his love of Beatrice.

To establish the scholastic context of the issues of passion and volition raised in *Epistola* 3, as well as in the episodes of the *Commedia* addressed later, this chapter engages the work of Peter of Trabibus, who was the *lector* in theology at Santa Croce at least between 1294 and 1296, giving lectures on the *Sentences* and holding public quodlibetal debates that Dante could have attended as a lay person. Manuscript evidence indicates that Peter was teaching at Santa Croce at the approximate time Dante claims in the *Convivio* to have undertaken thirty months of intensive philosophical study, frequenting the "scuole delli religiosi" and "disputazioni delli filosofanti" (*Conv.* 2.12.7; Dell'Oso 2022: 212–14). As a thinker, Peter was deeply influenced by Bonaventure and was likely a student of Peter John Olivi, who was *lector* at the Santa Croce *studium* between 1287 and 1289 (Dell'Oso 2022: 214–15). Trabibus occupies a notable place in the history of Franciscan thought as an important precursor to Duns Scotus, on both the rejection of the divine illumination theory of knowledge and on the idea of the formal distinction (Delorme 1924; Doyle 1984; Houser 1979; Huning 1968; Simoncioli 1956). If Dante did attend Trabibus' lectures on the *Sentences* and public quodlibetals, they would have provided him the opportunity to not only encounter but directly participate in debates about the most pressing philosophical, theological, and ethical issues of his day, furnishing him with a repertoire of scholastic methods and concepts, which as will be seen below, he readily and rigorously applies in his own work.

By approaching Dante's thought on passion and volition through Trabibus' *Quodlibetals* and *Sentences Commentary*, especially through an unedited question from Peter's Florentine *Ordinatio*—"utrum liberum arbitrium possit cogi," whether free will, or more literally, "free choice" can be compelled—this chapter contributes to a renewed focus in Dante studies on archival research regarding the poet's intellectual formation in his Florentine milieu during the 1290s (Davis 1965, 1988; Gentili 2020). In particular, I build off the work of Lorenzo Dell'Oso and Sylvain Piron, who have devoted studies to the relationship between Dante and Trabibus (Dell'Oso 2022; Piron 2008). My method for comparing Trabibus' work to *Epistola* 3 and the *Commedia* falls between the two. Without speculating on Dante's direct knowledge of Trabibus as does Piron, or surveying Dante's use of Aristotelian quotations as does Dell'Oso, I plumb the philosophical content of Trabibus' texts in order to highlight the depth of Dante's own philosophical and indeed scholastic approach to questions of passion, volition, and love.[9]

Appetition and Apprehension, and Generation and Corruption in *Epistola* 3: Dante's Innovation in the Natural Philosophy of the Passions

In the Third Epistle's account, a passion arises when an object causes a qualitative change (*alteratio*) in a subject (*subiectum*). Because the subject, in this case the concupiscible power (*potentia concupiscibilis*) or sensitive appetite, remains the same, one passion of love can transform into another only with respect to a different object: "de passione in passionem dico secundum eandem potentiam et obiecta diversa numero sed non species" (*Ep.* 3.2).[10] In other words, the new passion of love arises with respect to a different object of love, different not with respect to *species* or kind but with respect to *numero* or individual number or instance.[11]

Because love is seated in the sensitive soul and, therefore, governed by the natural philosophical principle "the corruption of one thing is the generation of another" (*generatio unius est corruptio alterius*), one passion of love can change into another (*Ep.* 3.4). Before the same appetitive power can undergo the same passion of love with respect to a different object, the first passion of love must corrupt or pass away. The corruption of the first passion implies the corruption of the object that generated it because the seat or subject of the first passion, remains. Once the previous object of love corrupts, the seat or subject of love, the *potentia concupisciblis*, can be "reserved for another" object that generates another passion within this same faculty of appetition (3.5). Thus, Dante concludes, yes, *de passione in passionem possit anima transformari*—the soul can be transformed from one passion to another. However, an essential aspect of Dante's argument is left implicit: how does a passion come to be or pass away with respect to an object in the first place?

In his commentary on the epistle, Marco Baglio does not account for the way an object activates the concupiscible power, and as a result, he does not explain the connection Dante makes between passion and generation and corruption. Informing virtually all scholastic accounts of the passions and underpinning the argument about the generation and the corruption of love in the Third Epistle is another Aristotelian principle: *appetitus sequitur apprehensionem*, appetition follows apprehension, as explained, among others, in Trabibus (Simoncioli 1956: 210), Aquinas (*ST* 1.79.1 ad. 2), and Dino del Garbo (Fenzi: 96). In short, this principle states that sense-perception is the efficient cause of passion. Dante's adherence to the notion that appetition follows apprehension is embedded in his characterization of the *potentia concupiscibilis* as a sensitive power (*Ep.* 3.5). The concupiscible power remains inactive unless brought into act ("in actum reducitur") by a passion that arises from an external object ("secundum...obiecta diversa," *Ep.* 3.5, 3.2). The passive verb form "reducitur" indicates how the concupiscible power is a passive potency and the seat of the *passio* of love: the *potentia concupiscibilis* needs to be acted upon by an object in order to operate. However, in the Aristotelian theory of perception to which Dante subscribes, here as well as in the *Commedia*, the sensitive appetite does not come into direct contact with the sensible object, only the senses do.[12] The sensitive appetite can only be acted upon by the object insofar as the object is apprehended or perceived: the object causes appetition via the medium of apprehension, *appetitus sequitur apprehensionem*.

For Aristotle and Dante, sensation, perception, and intellection follow a gradual process of dematerialization or abstraction. In Aristotle, the external senses are actualized, that is, brought into operation with respect to concrete objects. Sensation provides the raw material that actualizes the internal senses: the imaginative, cogitative, and memorative powers (Wolfson 1935). Imagination (*vis imaginativa*) is the internal sense that stores the sensible forms impressed on the soul via sensation (Pasnau 2001: 269). The cogitative power (*vis cogitativa*) discriminates these sensible forms as useful or dangerous, producing *intentiones*, rudimentary judgments about the appetibility of the object (Miner 2009: 72–82). The memorative power (*vis memorativa)* stores these *intentiones*, which are the basis of future judgments about similar objects (Miner 2009: 70, 79).[13] Physical sensation, sensible forms, and *intentiones*, albeit in different ways, actualize the sensitive appetite and, therefore, cause the passions of the soul (*passiones animae*) or emotions. (Miner 2009: 66–67). Just as the senses are moved by the objects of sense, the sensitive appetite has the disposition to be "naturally moved" by the various acts of sensitive apprehension (Miner 2009: 15).

In the major premise to the epistle's argument Dante employs this idea of a power's natural disposition to be moved. He repeats twice the phrase "naturaliter reservatur in alium" to describe the way the concupiscible power is able to be "naturally" actualized again if its first act corrupts: "Omnis namque potentia que post corruptionem unius actus non deperit, naturaliter reservatur in alium: ergo potentie sensitive, manente organo, per corruptionem unius actus non depereunt, et naturaliter reservantur in alium" (*Ep.* 3.5).[14] An intertext that has not previously been proposed for Dante's major premise is *Summa contra gentiles* 2.55.4, which states "in omni corruptione, remoto actu, manet potentia."[15] Lexically, Dante echoes Aquinas' verb *manere* ("manente organo"), and conceptually Dante accurately renders the idea that in every corruption, when the act is removed, the potency remains—it "is reserved for another." What the major premise affirms of sensitive powers in general, the minor premise applies to the *potentia concupiscibilis* in particular: if the seat of love, the concupiscible power, remains after the corruption of the passion that had first actualized it, that same power can be brought once more into actuality by another passion of love (*Ep.* 3.5).

The movement from potency to act, though, does not fully describe how an old passion passes away and a new one comes to be. Graziosi accurately notes, as does Baglio who follows her, that Dante understands the metaphysical change between act and potency to underpin the physical change of generation and corruption (Graziosi 1997: 62; Baglio 2016: 85–86). However, Graziosi and Baglio gloss over the precision of Dante's physics and metaphysics and fail to recognize the innovation of the *determinatio* in *Epistola* 3. Dante's insight is this: the generation and corruption of accidents/passions differ from that of substances. For both accidents and substances, corruption is a movement from act to potency, and generation is a movement from potency to act; however, when a *substance* generates or corrupts, the terms that comprise the act/potency dyad also change. In other words, the generation of a *tree* results from the corruption of a *seed*. The being of the seed corrupts, a movement from act to potency, as the being of the tree generates, a movement from potency to act. By contrast, the generation and corruption of accidents/passions do not fundamentally change the act/potency dyad that describes the substance itself: the substance itself remains. To continue the analogy, leaves on the tree generate and corrupt without the tree ceasing to exist. In short, when accidents/passions come to be or pass away, the substance undergoing those passions stays the same. As Dante writes, a passion can corrupt and another generate in its place because the potency "remains" and is "naturally reserved for another," that is, can be brought into act once again by another passion (*Ep.* 3.5). Thus, one passion can transform into another—*passione in passionem possit anima transformari* (*Ep.* 3.2). The application of generation and corruption to the transformation of one passion to another is the step in Dante's argument that not only marks *Epistola* 3's philosophical complexity but also its originality.

To fully appreciate Dante's philosophical intervention, it is necessary not only to distinguish between metaphysical and physical change but also the difference between two types of physical change at play in the epistle: generation and corruption, and alteration. Because "alteratio est transmutatio in passionibus," alteration is transformation in the passions, the type of change most properly said of the passions is not in fact corruption and generation but *alteratio* (L. 10.74, *De gen.* 319b 4). Alteration refers to the qualitative change between contraries that occurs as one instance of a single passion; *alteratio* is the change from warmth to cold, from illness to health, or from joy to sorrow.[16] Because alteration names the qualitative changes that passions produce in a substance, in a *quaestio* about change and passion, the expectation would be for Dante to discuss alteration and not generation and corruption. Baglio shares this expectation, which is the reason for his inaccurate gloss that "de passione in passionem," the transformation from one passion to another, refers only to a process of alteration (Baglio 2016: 83). *Epistola* 3 does not describe a change between contraries but the passing into non-being of one passion and the coming-into-being of another instance of that same passion: not the change from love to hate, but the corruption of one love and the generation of another. Dante's innovation is to go against the philosophical grain by applying generation and corruption not to substances but to passions themselves.

By treating this question of change and passion through the lens of generation and corruption, Dante goes further than his contemporaries. In his commentary on *Donna me prega* Dino del Garbo recognizes that Cavalcanti considers love to be a "created and generated thing" ("amor, de quo dictum est quod est res creata et generata") (Fenzi 1999: 96). But, as Barolini points out, Cavalcanti never addresses the question of an old love corrupting and a new love generating in its place (Barolini 2022: 225). Closer to Dante is Peter of Trabibus, whose Florentine quodlibetals and lectures on the *Sentences*, as mentioned, Dante could have attended at Santa Croce in the mid-1290s. In a question regarding the plurality of forms, a topic of heated and widespread scholastic debate in Dante's time, Trabibus makes a distinction between the generation and corruption of substances and of accidents (Huning 1968; Dales 1995; Gragnolati 2008).[17] Trabibus writes, "generatio substaniae unius corruptio substantiae alterius et generatio accidentis unius corruptio alterius accidentis," the generation of one substance is the corruption of another, and the generation of one accident is the corruption of another (quoted in Huning 1968: 145). Not only does he here cite the same Aristotelian principle of change as Dante, but Trabibus also maintains, as Dante does, that generation and corruption applies not just to substances but to accidents or passions as well. Taking what was said previously about how passions are a subset of accidents, Trabibus provides a vocabulary for the kind of change treated in *Epistola* 3. The soul's transformation from one love to another is an instance of relative or "accidental generation and corruption": the corruption of one accident/passion generates another. More will be said later about Dante and Trabibus, but it suffices now to say that Dante shares with him the notion of accidental generation and corruption. Dante goes beyond Trabibus, though, by applying the principle of generation and corruption not to accidents in the abstract, but to the passion of love and how it transforms.

In fact, Dante's further innovation in *Epistola* 3 is that he philosophizes transformation in love as a matter of both being and becoming, as a double-layered process of change involving both *alteratio* and generation and corruption. In a first-order sense, because every passion is an instance of alteration, one passion of love arising in the concupiscible power is a qualitative change of that faculty of desire. In a second-order and more dynamic sense, for one passion of love to transform into another, the first love and its *alteratio* must pass away before a second love and its new *alteratio* can come to be. "De passione in passionem" thus refers to a *process of change undergoing a process of change, alteration itself undergoing corruption and generation.* In the *Convivio*, Dante puts the relationship between alteration and generation and corruption as follows: "ogni cosa che si corrompe, sì si corrompe, precedente alcuna alterazione" (*Conv.* 4.10.9).[18] Enrico Proto seems to be the only *dantista* who has identified a possible source of this concept in Aquinas' opuscule *De mixtione elementorum*, which explains how alteration is "on the way" to generation and corruption: "unde alteratio est via ad generationem et corruptionem" (1915: 201–2). The idea that alteration leads to generation or corruption explains what Dante means when he writes that "the love of one can languish (*torpescere*) and finally perish, and insofar as the corruption of one thing is the generation of another, love for another may be formed again in the soul" (*Ep.* 3.4).[19] *Torpescere*, languishing or wasting away, is Dante's word for the alteration that precedes the corruption of love. How this languishing occurs is not addressed directly in *Epistola* 3. It is possible, however, to reconstruct a Dantean account given what the letter says about the relationship between apprehension and appetition, that is, how the passion of love arises with respect to a particular object or individual.

Languishing Love: Cecco d'Ascoli on the Corruption of the "Cuore," Dante on the Corruption of the Beloved

In the first chapter of the third book to his *Acerba*, Cecco d'Ascoli cites by name the sonnet accompanying *Epistola* 3, *Io sono stato*, and critiques its account of how one passion of love may become another by declaring that only death can separate lovers ("Non si diparte altro che per morte," 3.1.61). Here, I highlight a previously unremarked connection between Cecco's account of love and *Epistola* 3's assertion that a passion of love alters before it corrupts, that it can languish before it dies ("posse torpescere... denique interire," *Ep.* 3.4). After establishing the role of Venus' celestial influence in the genesis of love, and how love can arise from the interiorized image of the lady even in her absence, Cecco addresses how "the heart languishes" ("dico come il cuore langue") and the passion of love corrupts (*Acerba* 3.1.90).[20] According to Cecco, like the light that illumines the air, so too is the object of love seated in the "heart," which here signals a sensible power, perhaps the sensitive appetite (Pace 2020). However, just as an exceedingly bright visible object can overwhelm or corrupt the passive potency of the power of sight, so can love overwhelm the heart, making it languish (3.1.91–101). Cecco's account of languishing love differs in one important aspect from Dante's in *Epistola* 3. For Cecco, love languishes because the seat or subject of the passion corrupts. *Epistola* 3 describes precisely the opposite case: the sensitive appetite, the *potentia* that is the seat or subject of love, does not corrupt but remains intact. Rather, the corruption of the beloved causes the passion of love to languish and pass away.

Aquinas clarifies how the corruption of the passion of love in *Epistola* 3 can occur on the part of the object and not the subject of love. He enumerates three ways corruption occurs: by "the action of its contrary, the corruption of its subject, or the failing of its cause" (*SCG* 2.79.10). For Cecco, love corrupts due to the corruption of its subject: the faculty in which love is seated becomes completely overwhelmed. Dante, by contrast, states explicitly that the seat or subject of love, the concupiscible power, remains in potency to another passion. Therefore, in *Epistola* 3 the corruption of the first passion of love occurs by way of the corruption of the object that generated it, that is, *per defectum causa*. The motivation for Dante's innovation in *Epistola* 3—his application of existential change to qualitative change, that is, generation and corruption to alteration in the passions—is to explain philosophically how love corrupts (*interire*) when the beloved herself has corrupted, that is, has died (*Ep.* 3.4).

This does not necessarily imply, however, that Dante agrees with Cecco, that death divides lovers. In fact, Dante reaches the opposite conclusion via the same means. Dante indicates a temporal and processual dimension to the corruption of love: the passion languishes (*torpescere*) before it finally dies (*denique interire*) (*Ep.* 3.4). The implication is that the passion of love does not immediately extinguish upon the death (or corruption) of the beloved. As Dante attests in *L'amaro lagrimar,* a sonnet included in the *Vita nuova*, love persists death and the poet declares fidelity to the departed beloved, vowing to never forget her memory (Barolini 2022: 236–42, 264–65). Dante's poetic vow has a philosophical basis: because appetition follows apprehension, the passion of love can be generated not only with respect to the physical presence of the beloved but also in her absence, with respect to the remembered image of her perceived form. Like Cecco, Dante adheres to the notion that permeates the Italian lyric tradition: love arises as a function of the internalized image of the beloved, produced in the imagination and stored in the memory (Rea 2018: 121–22). But unlike Cecco, in *L'amaro lagrimar* Dante affirms that the passion of love can and ought to be kept alive by the poet's use of memory and imagination, in the very act of writing that calls up the remembered image of the departed beloved. In *Epistola* 3, then, a plausible explanation for how a love can languish before it finally perishes is that over time the image or memory of the departed beloved that sustains the passion of love withers in intensity and ultimately dies away. And once it does, a new passion, a new love, naturally (*naturaliter*), inevitably, generates in its place. In other words, the Third Epistle shows Dante using Aristotelian philosophy to make sense of the affective fallout from Beatrice's death, the gradual death of his love for her.

Beatrice and the Problems of Generation and Corruption, and Passion and Will in *Io sono stato*

Though she is not mentioned in the prose of the letter, Beatrice's death is the singular event, in both his biography and poetry, that forces Dante to confront the question of what remains of love when the beloved passes away. With regard to *Epistola* 3, this assertion does not violate the biographical fallacy, since the sonnet that accompanies the letter, *Io sono stato con amore insieme*, evokes Beatrice in its *incipit*: "Io sono stato con amore insieme / dalla circulazion del sol mia nona" (1–2).[21] As many have noted, the incipit is a deliberate echo of the beginning of *Vita nuova* 2, where Dante recollects his first encounter with Beatrice as a nine-year-old boy. The reference to the *Vita nuova* demonstrates how the sonnet begins by representing the moment of love's generation: the appearance of Beatrice ("apparuit iam beatitudo vestra") generates via apprehension the passion of love in the subject, Dante (*VN* 2.1). The sonnet ends by depicting the corruption of this love and the generation of another. More than this, *Io sono stato*'s reference to the *Vita nuova* shows that the corruption of the first passion love, in both the letter and the sonnet, refers to Beatrice and the impact of her death on Dante's love for her.

The image of the horse and rider, which follows the *incipit*'s evocation of Beatrice and the generation of love, carries through to the end of the poem, figuring the various elements of love-passion about which the letter speaks. The rider, personified *Amore*, pricking ("sprona," line 3) the horse, is an apt image for the activation of passion in the concupiscible power. As is the horse, so the sensitive appetitive is passive until brought into act. It is not a self-mover but a moved mover: just as the rider pricks the flank of the horse, which, in turn, makes the horse move, so the concupiscible power is stirred to actuality by *Amore*, the passion of love.

Reprising the image of the rider and the horse from the sonnet's opening, the last verses provide a poetic demonstration of *corruptio unius generatio sit alterius* (*Ep.* 3.4; Barolini 2022: 247–51):

> Ben può con nuovi spron punger lo fianco,
> e qual che sia'l piacer ch'ora n'addestra,
> seguitar si convien, se l'altro è stanco. (lines 12–14)[22]

If a former love-passion ("l'altro") is outworn ("stanco"), i.e. corrupted, then a new love with new spurs ("nuovi spron") can pierce the flank ("punger lo fianco") of the lover. That the horse's flank symbolizes the concupiscible power of the lover is also seen in the fact that what leads the horse ("ora n'addestra") is a "piacer," literally a pleasure. Recalling that the object of the concupiscible power is the pleasant good, "piacer" is not only poetically but philosophically apt. Moreover, just as the epistle alludes to the natural necessity by which the concupiscible power is once again brought into act, here in the sonnet's last line it is necessary that once spurred, the horse, the lover, follows this new love ("seguitar si convien"). The natural necessity involved in the generation and corruption of passion provides an explanation as to why in the sonnet *Io sono stato* declares that in Love's arena, free will was never free ("Però nel cerchio de la sua palestra / liber arbitrio già mai non fu franco," lines 9–10; Barolini 2022: 250). If Love's *palestra* is the arena of natural necessity, then love is not a matter of choice, but of physics.

While the *Commedia* departs from *Epistola* 3's and *Io sono stato*'s account of the relation between passion and volition, it remains consistent with their phenomenology of apprehension and appetition. As in the letter and sonnet, Vergil's discourse on love in *Purgatorio* 18 maintains that love arises with respect to an object via apprehension: "vostra apprensiva da esser verace / tragge intenzione" (lines 22–23).[23] Once apprehended, this *intentio* unfolds within and the soul turns toward it in the act of appetition (lines 23–24).[24] *Purgatorio* 18 gives a vernacularized account of *appetitus sequitur apprehensionem*. The difference between this canto's account of love and *Epistola 3*'s and *Io sono stato's* is that in the *Purgatorio* the will holds the threshold of consent to the passion of love ("de l'assenso de' tener la soglia," *Purg*. 18.63). That is, the human being has the freedom to choose to follow the inclination of its desire or not. Even if, says Vergil, we took the thesis of the Third Epistle to be true, that every passion of love arises as a matter of natural necessity—"poniam che di necessitate / surga ogne amor che dentro a voi s'accende" (*Purg*. 18.70–71)—the power to restrain that love, via free choice (*liberum arbitrium*), would remain ("di ritenerlo è in voi la podestate" 18.72). Even if appetition necessarily follows on apprehension, the will, the faculty of rational appetition, supervenes on sensitive appetition and neutralizes natural necessity. Hence in the *Monarchia* Dante defines the judgment (*iudicium*) of the *liberum arbitrium*, as the "medium between apprehension and appetition."[25] In *Epistola* 3, however, Dante characterizes *appetitus sequitur apprehensionem* as a matter of natural necessity, of Aristotelian physics rather than as choice. This difference marks the crucial role the will plays in Dante's understanding of love in the *Commedia*, for in all three *cantiche* Dante conceives the will as radically free. As will be seen below, the nature of the will and its freedom are central to Dante's articulation of the difference between love as an affect of the will and as a passion of the sensitive appetite.

Examples of Intellectual Love Before the *Commedia*: *Doglia mi reca* and *Amor che nella mente*

The fact that in *Epistola* 3 Dante treats love as a function of sensitive desire is surprising, if not jarring, given the emphasis on the intellectual nature of love in the *Vita nuova* and in the *Convivio*. To situate the philosophical distinctiveness of the epistle's notion of love-passion, as well as to differentiate the *Commedia*'s articulation of love as an affect of the will from the notions of intellectual love that Dante explored previously, it will be useful to briefly discuss two canzoni, *Doglia mi reca* and *Amor che nella mente mi ragiona*. In these canzoni, Dante intellectualizes love in two different ways: the former ethically, the latter metaphysically. In *Doglia mi reca* love is a virtuous expression of desire when passion is regulated by reason. In *Amor che nella mente* love is no longer a passion but an act of intellectual apprehension, or more precisely, the failed apprehension of an angelized beloved who exceeds the human intellect.

In *Doglia mi reca* Dante's focus is on the morality of love, on love inside and outside the "orto di ragione" (line 147).[26] Inside reason's garden, love is virtuous; outside of it, love is better called lust, a bestial appetite, "chiamando amore appetito di fera" (line 143; Barolini 2006: 47–69). In *Doglia mi reca* Dante envisions the optimistic possibility of sensitive appetition in harmony with reason, which contrasts *Io sono stato's* account of love that falls outside the "palestra" or arena of free choice, the composite power of reason and will. In *Io sono stato* the "consiglio" that issues from the "libero arbitrio" is like an arrow that fails to hit its target, useless to control passion.[27] *Doglia mi reca*, however, not only shows the possibility of the rational regulation of passion, but asserts it as a moral imperative. The second stanza's discourse on virtue articulates the right gradation of love: romantic passion governed by reason is virtuous; "vertute" finds its measure ("misura") in heaven ("ciel"); and the *fattor* of virtue, *Amore*, honors "vertute" in the "beata corte" of paradise (lines 27–42; Foster and Boyde 1967: 2.301–3). In other words, the canzone treats romantic love like a secondary good that "becomes an ethical path to the first good," to God (Gentili 2005: 192).[28] In *Doglia mi reca*, passion that is subject to reason is a stepping stone to virtue and even divine *Amore*.

If *Doglia mi reca* imagines a healthy rapport between reason and love, *Amor che nella mente mi ragiona* conceives love as the very operation of reason or intellect. The canzone begins with love reasoning or speaking of the beloved directly within the mind of the poet. The *incipit's* collocation of "mente" and "ragionare" indicates how the canzone intellectualizes both the seat of love and its operation. In fact, love's speech is so exalted and purely intellectual that it is ultimately ineffable and incomprehensible to the poet, surpassing "nostro intelletto" and "parlar nostro" (lines 10–12, 16–18). In addition to this inexpressibility *topos*, Dante develops a poetic of metaphor and being that surpasses the poetic of similitude found in his predecessor Guinizzelli.[29] In Guinizzelli's *Al cor gentil rempaira sempre amore* the *donna* merely resembles an angel in her "angelica sembianza," while Dante's canzone fashions the *donna* as ontologically on par with an angelic intelligence. In fact, as they contemplate

God so do the angels contemplate the *donna*, and like the angels themselves the *donna* receives grace or the "virtù divina" (lines 37–38) beyond the demand of human nature ("oltre 'l dimando di nostra natura," line 29). In the *Convivio* commentary, Dante asserts that in the same way, the angelic intelligences receive the emanation of divine light directly from God, without mediation ("sanza mezzo"), so too does the *donna*.[30] As will be seen in the *Paradiso*, already here in *Amor che nella mente* Dante conceives of an unmediated relationship of the soul to God: the *donna* is so pleasing to God that she receives a ceaseless infusion of divine grace ("Suo esser tanto a Quei che lel dà piace, / che'nfonde sempre in lei la sua vertute," lines 27–28). However, in the *Paradiso* divine love infuses the human will, uniting it to God, whereas in the canzone, love remains within the province of intellect or "mente."

The *Convivio* commentary amplifies the intellectualizing tendency of the canzone to a degree that goes beyond the poem's letter. Drawing on the *Liber de causis* and the Neoplatonic doctrine of the divine idea, Dante transfigures the *donna* into nothing less than the "intentional exemplar of the essence human form that resides in the mind of God."[31] Just as the forms of what an artisan intends to make reside in her mind, so too the exemplar forms of all created things reside in the divine mind (Doolan 2008: 1; Cullen 2006: 71–87; McIntosh 2021). According to the *Convivio*'s gloss, the *donna* of *Amor che nella mente* is the "perfect form of the human being" known by God apart from material instantiation.[32] While she is known perfectly in the divine mind, the *donna*, the exemplar form of humanity, is imperfectly manifest (because made material) in each individual.[33] Whereas in the canzone the *donna* remains embodied, in the prose commentary, Dante fully dematerializes the *donna* making the object of love an object of pure intelligibility, a divine idea.[34] In maximally intellectualizing love and the beloved, Dante understands the project of his commentary to *Amor che nella mente* to be a defensive one, to ward off any "falsa oppinione" that in his poetry love signifies "sensibile dilettazione," the pleasure of passion (*Conv.* 3.3.12; Barolini 2022: 253).

In this light, it is striking how *Epistola* 3 and *Io sono stato* prove these "false opinions" true: the letter and the sonnet undeniably treat love as a passion of sensual desire, poetically and philosophically. It might be said, then, that despite his own efforts to reduce desire to intellect, for Dante desire is an irreducible element of love. In fact, the interplay between poem and commentary in *Amor che nella mente* attests to this. In the canzone, love speaks of the beloved within the poet's "mente" but does so "disiosamente," desirously (line 2). Here the poem seems to characterize "mente" holistically, as including both intellect and desire. However, in the *Convivio* Dante overdetermines the meaning of "mente" as intellectual. On Aristotle's authority, Dante defines *mente* as the "highest and most noble part of the soul" whose "ultima potenza" or highest faculty is "ragione" or reason (*Conv.* 3.2.14–16). The canzone, however, resists the intellectualizing procedure of the commentary and retains the desirous dimension of love, indexing Dante's persistence as a poet of desire.

In the *Commedia* love remains a function of desire, but given the poem's radical commitment to the will's freedom, love can no longer be a passion outside the bounds of reason or volition. Hence the critique of lust in *Inferno* 5 and Vergil's ethically inflected phenomenology of love in *Purgatorio* 17 and 18. Yet given the *Commedia*'s trajectory toward the knowledge and love of God, love cannot remain a passion at all, even inside the bounds of reason and will. The pilgrim's unmediated experience of divine love transcends the limits of human intellect and desire. Therefore, an overarching poetic and philosophical concern emerges especially in the latter half of the *Purgatorio* and the whole of the *Paradiso*: the reconciliation of human desire with the love of God. If God is pure act, whose essence is his existence, beyond the bounds of human reason, in what, then, from the human side of things can the love of God consist? Below, we shall see how in the *Commedia* Dante returns to the notion of generation and corruption to articulate love as a spiritual reality infusing the will.

Generation, Corruption, and the Creation of the Soul in *Paradiso* 7

Paradiso 7 expands *Purgatorio* 25's account of God's direct creation of the rational soul. The soul's unmediated creation means that in the cosmic chain of being the soul stands above all corruptible things and is, therefore, completely free; the soul cannot be causally determined by any created thing. In other words, in *Paradiso* 7 Dante elaborates how the soul's ontology implies the freedom of its will and its direct relationship to God. In turn, the soul's unmediated relationship to God makes possible the loving union of the human will with the divine.

In *Paradiso* 7, the only canto in the *Commedia* in which the noun "corruzione" appears (and it does so twice, in consecutive *terzine*),

Beatrice uses the Aristotelian principle of generation and corruption to articulate the ontology of the soul and the difference between mediated and unmediated creation. Beatrice intuits the *dubbio* that clouds the pilgrim's mind:

> Tu dici: "Io veggio l'acqua, io veggio il foco,
> l'aere e la terra e tutte lor misture
> venire a corruzione, e durar poco;
> e queste cose pur furon creature;
> per che, se ciò ch'è detto è stato vero,
> esser dovrien da corruzion sicure." (*Par.* 7.124–29)[35]

The source of Dante's doubt is his faulty syllogism: from the major premise that God creates all things, and the minor premise that God himself is incorruptible, the pilgrim reasons that what God creates must also be "da corruzion sicure" (line 129). The issue here is that at this point in his journey through paradise, Dante has not yet grasped the implications of secondary causality. Beatrice responds that even though all things are created by God, the elements and the souls of animals and plants are created in an indirect way through the secondary causes of nature, the "creata virtù" of the celestial spheres (lines 135–38). Being created through created things, the elements, and vegetative and animal souls are, therefore, transient and mutable, subject to generation and corruption (*Par.* 7.139–41). One of the lessons of *Paradiso* 7 is that indirect creation, via secondary causes, implies corruptibility, whereas direct creation, by God, implies incorruptibility.

That which is created without mediation, "sanza mezzo," is not only incorruptible but also completely free:

> Ciò che da lei sanza mezzo distilla
> non ha poi fine, perché non si move
> la sua imprenta quand' ella sigilla.
> Ciò che da essa sanza mezzo piove
> libero è tutto, perché non soggiace
> a la virtute de le cose nove. (*Par.* 7.67–72)[36]

Creation "sanza mezzo" grounds the soul's immortality and incorruptibility ("non ha poi fine," line 68) and its freedom (line 71). Like the angels and celestial spheres (line 130), God himself directly creates the soul, which grants the soul a position in the order of being that renders it incorruptible and completely free ("libero è tutto," line 71) from the causal power ("virtute") of corruptible things or "cose nove" (lines 71–72). In *Epistola* 3 and *Io sono stato* Dante had used the principle of generation and corruption to mark the will's unfreedom. But here in *Paradiso* 7, Dante applies the same Aristotelian principle to mark its freedom: the soul's direct creation entails its incorruptibility, which, in turn, entails the freedom of its will. Dante also recognizes a further implication: because the will, just as the intellect, transcends matter, the love which infuses the will also transcends material passion. The direct causal relationship between Creator and creature imbues the soul with the aim of its desire, the "somma beninanza," God:

> Ma vostra vita sanza mezzo spira
> la somma beninanza, e la innamora
> di sé sì che poi sempre la disira. (*Par.* 7.142–44)[37]

Breathed forth directly from God, the human soul is ontologically different, yet not separate from the ground of its being in God. The soul's nature is to always tend toward and desire this ground, in order to, as Dante says in the *Convivio*, more fully actualize or "fortify" its being (*Conv.* 3.2.7). Its direct creation and immateriality are the ontological bases for the soul's passionless love of God.

The intellectual backdrop to the issues of creation, freedom, and love that preoccupy *Paradiso* 7 is the Condemnations of 1277 (Hissette 1977; Wippel 1977, 1995). Drafted by the masters at the University of Paris, under the auspices of the bishop of Paris, Stephen Tempier, the Condemnations were a series of 219 theological and philosophical propositions addressing issues in Aristotelian philosophy and various kinds of determinism deemed incompatible with Christian doctrine, nineteen of which specifically consider matters of the will. The work of Peter of Trabibus attests to how the fallout from the Condemnations made its way into the Florentine milieu that shaped Dante's philosophical development. In the *Ordinatio* of his *Sentences Commentary*—a fully revised version of the *Lectura*, or partially revised redaction of the courses on the *Sentences* he gave at Santa Croce between 1294 and 1296—Peter of Trabibus evokes several propositions from the Commendations by addressing the question: *utrum liberum arbitrium possit cogi*, whether free choice, the composite power of reason and will, can be compelled (Florence,

Biblioteca Nazionale Centrale, MS Conventi Soppressi B.V.1149, 127vb-128ra; Piron 2008: 80–1; Friedman, forthcoming).[38] Trabibus' response is notable for its similarities to the way Dante grounds the will's freedom with respect to the will's position in the order of creation. Like Dante in the *Commedia*, Trabibus considers the will to have a radical freedom; it cannot be compelled at all, "neither by a created nor uncreated agent," neither by God nor anything within nature.[39] This is because there is no created power that exceeds the will ("nulla virtus maxime creata est quo ad hoc est superior eo," 128ra). Trabibus' "virtus creata" finds its vernacular equivalent in Dante's "creata virtù," the created power of the celestial spheres and angelic intelligences that gives form ("informante") to that which generates and corrupts (*Par.* 7.135–37; Bemrose 1983: 95–96). Conceptually, Dante aligns with Trabibus' notion that no other created power exceeds the will, since it is not subject to created agents ("non soggiace / a la virtute de le cose nove," *Par.* 7.71–72). However, what distinguishes Dante from Trabibus on the position of the will in the order of creation is Dante's recourse to Aristotle's notion of generation and corruption. For Dante, the soul's direct creation means that it does not come to corruption, and because the soul is incorruptible, its will is completely free, undetermined by the causal influence of corruptible things.

Trabibus does align with Dante on the way the soul's ontology implies the will has an unmediated relationship with God. In a public quodlibetal debate Dante could have attended at Santa Croce, likely in the spring of 1295, Peter discussed the question *utrum liberum arbitrium subsit corporibus coelestibus*, whether free choice is subject to the influence of the celestial bodies (Florence, Biblioteca Nazionale Centrale, MS Conventi Soppressi D. VI. 359, fol.108vb-109ra; Simoncioli 1956: 230–2; Piron 2008: 77–8). In his text, Peter describes how God may be said to move the will as its "principle" (*principium*) (Simoncioli 1956: 231). Because the will's act is that of a created being, God can be said to cause the will's act both "mediately" (*mediate*) and "immediately" (*immediate*): with mediation insofar as God sustains the being of the will, which moves itself to elicit its free act, and without mediation because God, as the first cause of all being, is more primary or "intimate" (*intimius*) than all other causes (Simoncioli 1956: 231).

Trabibus characterizes another way the soul has an unmediated relationship to God: while he does not compel the will, God can transform the will's affect—"affectio voluntati subtrahatur et contraria immutatur"—as when he makes a lover of temporal things a lover of things eternal, "solus Deus ut facit cum de amatore mundanorum facit amatorem eternorum" (MS Conv. Soppr. B.V.1149, 128ra). This notion that God can change the will from one affect to another—and indeed from *one kind of love to another*—has direct resonance with the problem of change and the passions in *Epistola* 3. But there is a crucial difference in the vocabulary of love between the epistle and the *Purgatorio*, *Paradiso*, and Trabibus: the difference between *passio* seated in the sensitive appetite and *affectio voluntati*, which can refer to a category of affect beyond passion and seated solely in the will (MS Conv. Soppr. B.V.1149, 128ra). Dante's depiction of Statius' liberation from purgatory reflects both Trabibus' notion that God can influence the affects of the will and that the will has passionless affects. In other words, Statius' liberation is evidence for a Dantean account of the experience of spiritual love and joy that belongs solely to the rational soul, distinct from material passion and sensitive appetition.

The Joy of the Will and the Paradoxes of Passion: Statius' Liberation and Dante's Smile

The entire mountain of purgatory quakes like it is about to collapse, and a cry rings out through the second realm. Though frozen with the fear of "one who goes to meet his death," the pilgrim has never desired to know anything more than the cause of this earthquake (*Purg*. 20.145–47). Dante learns of it from Statius, who encounters Dante and Vergil like the newly risen Christ who meets the two travelers on the road to Emmaus (*Purg*. 20.14–56; 21.7–9). The seismic shocks, explains Statius, have not a physical but a metaphysical cause, the liberation of his own soul (Barolini 2014):

> Libero è qui da ogne alterazione:
> di quel che 'l ciel da sé in sé riceve
> esser ci puote, e non d'altro, cagione…
> Tremaci quando alcuna anima monda
> sentesi, sì che surga o che si mova
> per salir sù; e tal grido seconda." (*Purg*. 21.43–60)[40]

"Alterazione," a *hapax legomenon* in the *Commedia*, is a freighted term. As discussed above, in Aristotelian natural philosophy *alteratio* is the name for the qualitative change proper to the passions: "alteratio est transmutatio in passionibus" (L. 10.74, *De gen.* 319b 4). To describe the physics of the mountain Statius uses "alterazione" in its broadest sense, referring to the qualitative changes or passions of the elements.[41] There is no weather—no rain, no hail, no wind nor snow—in the regions above purgatory's gate because they are free from all *alteratio* (lines 43–54). This means that only what is incorruptible, that which is from and of heaven ("di quel che 'l ciel da sé in sé riceve," line 44), can be the cause ("cagione" line 45) of change in the mountain's upper regions (Chiavacci Leonardi 2016: 2.621). The phrase "sé in sé riceve," which applies to the celestial spheres and their angelic movers, mirrors the one Statius uses four canti later to describe the rational soul: "sé in sé rigira" (*Purg.* 25.75). This lexical echo emphasizes, as we also saw in *Paradiso* 7, how the angels, celestial spheres, and human souls share ontological features: they are directly created by God and are, therefore, incorruptible and free from causal determination by corruptible things, those which bring about "alterazione." In anticipation of the pilgrim's own *disio* and *velle* being moved by divine love and circulating like the angels and celestial spheres, the motion of Statius' will, liberated through *divina giustizia*, acquires something like the force of celestial motion, causing the mountain to tremble. As the *beati motori* move the celestial spheres, so Statius' freed will moves purgatory's mountain as an immaterial cause producing material effects.

But there remains an important difference: like an angelic intelligence the soul is created "sanza mezzo" and incorruptible, but unlike an angel the human alone is the link, the *mezzo*, between the corruptible and incorruptible: "homo solus in entibus tenet medium corruptibilium et incorruptibilium (*Mon.* 3.15.3–4; Moevs 2005: 156–57). For Dante, the human is the *medium* between the corruptible and incorruptible because it unites the spiritual and material within its own being, as body and soul. And although in purgatory the soul is separated from its physical body, it enlivens a virtual body capable of sensation and passion (*Purg.* 25.88–99). Hence the process of the soul's purgation is a dynamic between the immaterial will and material passion. The linguistic and conceptual pressure Dante places on the idea of *alteratio* points up this dynamic: "alterazione" applies as much to the passions of the elements as it does to the emotional passions of the soul. It follows, then, that just as the mountain is free from the *alteratio* of the elements, so too in purgatory the will is made free from the *alteratio* of the passions (Barolini 2014).

As he does in *Paradiso* 7, in *Purgatorio* 21 Dante uses Aristotle to go beyond Aristotle: the absence of Aristotelian *alteratio* in the upper regions of purgatory throws into relief how Statius' joy is not a material passion but a spiritual affect of a will made ready to be in the presence of what it loves most, its ultimate Good, the love of God.

The experience of liberation, Statius says, is the soul's spontaneous enjoyment in feeling its will purified and free to ascend to paradise:

> De la mondizia sol voler fa prova,
> che, tutto libero a mutar convento,
> l'alma sorprende, e di voler le giova. (*Purg*. 21.61–63)[42]

The will is the only proof of the soul's purity ("de la mondizia sol voler fa prova," line 61), and the soul rejoices in its purified will ("di voler le giova," line 63). Unlike a passion of the sensitive appetite, this joy is an *affectio voluntati*, as Trabibus terms it, an affect of the will. And as Aquinas explains, a "spiritual joy" (*spirituale gaudium*) has its seat in the will, and indicates the presence and rest (*quies*) of the will in what it loves, which for the purgatorial soul is nothing other than God and his love ("quod de Deo habetur, ex caritate causatur," *ST* 2.2, q.28, a.1). Bonnie Kent notes that "we often find scholastics attributing *gaudium* to the will" because the joy experienced by the soul in the afterlife has "no *necessary* connection to the body" (Kent 1995: 211, my emphasis). In other words, the joy of the will differs from the joy that is a passion of the sensitive appetite because the cause of the former is universal and spiritual, the love of God, whereas the cause of the latter is particular and material, an object of sense. Dante's purgatory, by Kent's measure, is thoroughly scholastic: because there is no "alterazione" on the higher points of the mountain, and since the passions of the soul are alterations, Statius' joy cannot be a passion (*SCG* 2.81.15). Just as the *alteratio* in the elements cannot cause change at this level of purgatory, so too *alteratio* in the passions cannot cause the spiritual affects of the will in the liberated soul.

The difference between sensitive desire and volition is also dramatized in the way Statius characterizes the inner conflict the soul feels between its passions and will while undergoing purgation, before liberation:

> Prima vuol ben, ma non lascia il talento
> che divina giustizia, contra voglia,

come fu al peccar, pone al tormento. (*Purg*. 21.64–66)[43]

Placed in the rhyme positions of the *terzina*, the soul's desire for the "tormento" of atonement formally and conceptually mirrors the "talento" or inclination to sin it had in earthly life, that is, the inclination of the will to turn to particular goods less than God. *Talento* here should be taken as a deliberate echo of *Inferno* 5, where Paolo and Francesca and the other "peccator carnali" are damned for subjecting reason to desire, "che la ragion sommetono al talento" (*Inf.* 5.37–39). *Talento* thus signifies how sensitive appetition may incline or divert the rational appetite. For both Dante and Trabibus, what allows the sensitive appetite to exert influence upon the will is that both sensitive and intellective appetition are faculties of desire rooted (*radicati*) in the one and same essence of the soul; hence they are intertwined to the extent that if the sensitive appetite is drawn to something, so is the will.[44] The souls in the second circle of hell epitomize disorder and disharmony between passion and will: their *talento* permanently diverts their will away from God, and even obscures the fact that their wills are free. Yet in the *Commedia*, passions are not evil in themselves; rather, they acquire a moral valence insofar as they play a role in the will's turning away from or to the good, to God. For Dante, it is not what we feel—fear or desire, sadness or joy—but the way we will, in response to what we feel, that deserves damnation or merits salvation (*Purg*. 18.61–75). Because the very nature of the human is to be "the *medium* between the corruptible and incorruptible," in purgatory, the soul must learn to harmonize the material and spiritual dimensions of its desire, its passions and its will (*Mon*. 3.15.3–4). Hence for the souls in purgatory desire functions paradoxically: now an impediment, now an aid to liberation.

Upon entering purgatory, having confessed and with hope of eventual salvation, the soul wills the Good ("prima vuol ben," *Purg*. 21.64). Yet against this will ("contra voglia," line 65) the soul desires the *tormento* of its penance and the delay of its entry into paradise. Penitential *talento* inverts the soul's earthly *talento*: whereas on earth, the soul's wayward desire worked against God's will for the soul to turn to God, in purgatory the soul's desire works against its *own* will to return to God. The desire for penance in purgatory, given to it by *divina giustizia* is in direct proportion to the soul's desire on earth to "turn [its] steps" away from the highest Good and "follow the image of false goods," the aim of its passions (*Purg*. 30.130–31). The mechanism of purgatory, then, is not only the restoration of time, "dove tempo per tempo si ristora," but also the restoration of desire, where *talento per talento si ristora* (*Purg*. 23.84).

The purgatorial soul's desire and will are opposed only until the soul is surprised ("sorprende") to find they are not and never were in conflict (*Purg*. 21.65). The more the soul conforms its will to God's, assenting to the *talento* that is contrary to its own will ("contra voglia") but instilled in it by *divina giustizia*, the closer the soul actually is to realizing its will for paradise. Hence Forese Donati says, "io dico pena, e dovria dir sollazzo" (*Purg*. 23.72). The pain of purification is a consolation when it is experienced as ordered to the will's ultimate Good in God. And as Aquinas tells us, it is not unfitting that the *gaudium spirituale* that regards the love of God is compatible with an admixture of *tristitia*, for "we grieve over what opposes this participation in the divine Good in ourselves" (*ST* 2.2, q.28, a.2).[45] Whereas in the second circle, the damned souls let passion permanently turn their will from God, and they sorrow over their unwilled suffering, in the upper regions of purgatory souls willingly endure their *tormento* in accord with divine justice, and sorrow only over their own separateness from God. The journey of the soul in purgatory, at first its suffering and later its rejoicing in the process of purification, is Dante's poetic dramatization of the *mutatio* or change in the affect and orientation of the soul's will—from once a lover of worldly things to now a lover of things eternal—a change which, as Trabibus says, only God can bring about, or as Dante says, only *divina giustizia*. One remains an "amatore mundanorum" as long as one prefers one's own will over God's.

Paradoxically, then, even though the soul wills to ascend to paradise, it remains in purgatory only so long as it has the desire to remain there; and, by adhering to this desire to stay in the second realm, the soul is purified and its will made free for paradise. Statius himself enacts this idea when, after the emancipation of his own soul, he lingers with Vergil (and Dante). Statius says he would have gladly spent more time in purgatory if only he could have shared earthly life with Vergil (*Purg*. 21.100–02). In the irony of the narrative, Statius makes his counterfactual desire a fact: by staying with Vergil and Dante rather than rushing headlong to paradise, Statius shares with Vergil some of the afterlife instead of the earthly life. Perhaps counterintuitively, but not contradictorily, Statius' desire to remain in purgatory and his will to ascend heavenward are, quite literally, no longer opposed: Vergil and Dante are also walking the path to paradise. Of course, Statius will only ascend with the pilgrim, but he walks with Vergil out of love.

Hence Dante cannot hide his smile from Statius. The pilgrim knows Statius unknowingly stands before the *poeta* he speaks of and reveres, whom he owes both his poetry and conversion—Vergil (*Purg*. 21.109–23; *Purg*. 22.73–75; Barolini 2014). Witnessing and participating in this moment of recognition, friendship, and fellowship between Statius and Vergil, the pilgrim's "sorris[o]" (line 109) reveals him as one of the "veraci" (line 108) whose smile follows more the passion from which it springs ("a la passion di che...si spicca," line 107), and less the will ("men seguon voler," line 108) that would restrain it. The pilgrim's joyful "passione" is so intense that it exceeds his "voler," his will to obey Vergil's command to remain silent about his identity to Statius; the will cannot do all ("non può tutto la virtù che vuole," line 105). But the pilgrim here does not replay Francesca, submitting reason and will to passion. Rather, Vergil releases Dante from the command of silence because he recognizes it would be wrong to stifle the pilgrim's smile, which springs from a passion of joy that obeys a higher authority than Vergil's or the pilgrim's own will: Dante shares in Statius's selfless (and self-effacing) love for Vergil himself. Hence Dante's "passione" is "verace," is true, borne by a love that leads beyond all passion.

Dante's smile and Statius' joy demonstrate how the the purpose of the soul's time and experience in purgatory is to undergo a re-education of the will and the passions. The soul needs to learn to harmonize earthly and spiritual desire within itself, and in the process, harmonize its own will with that of *divina giustizia*. Dante's failure to hide his smile shows how rightly ordered passion, an emotion that springs from and toward rightly ordered love, can aid the will on its way home to paradise. Statius' lingering with Vergil and Dante reveals that once the will is purified, spiritual joy is not hindered by delighting in ties to the world, especially to those whom we feel closest. When it feels it is free, what the soul wants and what the soul wills no longer work to opposing ends because both are directed to the fulfillment found in their final end, in God, in love.

St. John's Examination on Love in *Paradiso 26*: Passion Transfigured

If Statius' liberation demonstrates a Dantean account of joy as an affect of the will purged from passion and turned to God, the examination on love St. John gives the pilgrim in *Paradiso* 26 demonstrates how Dante also considers love an affect of the will beyond passion. Moreover, *Paradiso* 26 is a canto wherein, after being tested by St. John, the pilgrim meets Adam, who rewrites not only the account of Adamic language given in the *De vulgari eloquentia* but also that of medieval tradition writ-large. In the *Commedia* language is no longer originally immutable and divinely co-created, and only later dispersed and made mutable at Babel; according to Dante's Adam, the language of the first man was always already a vernacular, an "opera naturale" that evolves with the preferences of speakers and passes into and out of existence (*Par. 26*.124–32; Barolini 2022: 162–79). In the examination on love, then, Dante demonstrates his poetic ability to transform his own vernacular into a vehicle for the expression of divine love, illustrating how for Dante language itself is an "immanent analogue of the Incarnation" (Mazzotta 1972: 6). All vernaculars, depending on their use, are potentially incarnational and can convey something of the experiential reality of the love of God—including, as Dante shows in *Paradiso* 26, the lyric language of passion.[46]

St. John first asks Dante to declare the aim of his soul, to which Dante gives an intellectual response: through philosophical arguments ("filosofici argomenti," *Par. 26*.25) and the authority ("autorità," line 26) of scripture, Dante says he has learned to love God as the highest Good. The good, insofar as it is understood or known as good, kindles love (*Par. 26*.28–29). Therefore, the pilgrim continues, if one understands the truth, then one ought to love that "essenza" who is the source of all good, since each good that lies outside this essence is merely its reflection (*Par. 26*.31–36). Here Dante reprises the principle that appetition follows intellection: through human intellect and scriptural authorities ("intelletto umano / e per autoritadi," *Par. 26*.46–47), Dante's love of God follows on his understanding of God as the *summum bonum*. The principle *appetitus sequitur apprehensionem* applies just as much to *caritas* and the will as it does to the passion and the sensitive appetite.[47]

However, St. John is dissatisfied with Dante's answer:

> Ma dì ancor se tu senti altre corde

tirarti verso lui, sì che tu suone
con quanti denti questo amor ti morde. (*Par.* *26*.49–51)[48]

And Dante responds:

[...] "Tutti quei morsi
che posson far lo cor volgere a Dio,
a la mia caritate son concorsi. (*Par.* *26*.55–57)[49]

The language of John's question and Dante's response is notable for the way it imports the violent and sensual lexicon of the *Rime petrose*, only to transfigure that lexicon into a spiritual register. As Barolini has shown, the "morsi" or bites of love recall Dante's canzone *Così nel mio parlar voglio esser aspro*, where he writes "la morte, con ogni senso / co li denti d'Amor già mi manduca" (lines 31–32; Barolini 2022: 97-8).[50] Love is so overwhelming it is tantamount to a force of death masticating the senses. The canzone's physical and violent language, that "each sense" is bitten by the "teeth of love," casts love as a sense-based and embodied passion. More than this, the very image of the "bite" of love opens onto a long tradition of discourse on the passions that has roots in the Stoics and is transmitted to the Middle Ages through Cicero and Augustine. For the Stoics, *praepassiones* or preliminary passions are involuntary feelings that stir even the sage's soul, before he can consent to the feeling or not; the Stoics describe these *praepassiones* as "bites" (*morsus*) or "little contractions" (*contractiunculae*) of the soul (*Tusc. Disp.* 3.82–83; Graver 2007; Byers 2012: 100–26). The "bites" of the preliminary passions are unwilled and non-consensual, just as in *Così nel mio parlar* love is described with the language of sensation, passion, and compulsion. However, the teeth of love about which St. John speaks refer to the love of God, not to a passion but a spiritual affect that inheres within a completely free will. Rather than the intellectual or apprehensive dimension, St. John urges Dante to sound ("suone," *Par.* *26*.50) the felt, experiential dimension of divine love. Perhaps counterintuitvely, the imaginative and linguistic resources of the erotic and violent lyric fittingly convey the intensity of spiritual love. The vernacular language of passion, sensation, and the body is the language of manifestation, tangibility, and visibility, allowing Dante to meet St. John's demand to express incarnationally the reality of spiritual love through the language of the flesh, to sound outwardly the inward love of God.[51]

The image of the cords of love that draw the soul to God is another inversion of a lyric formula: the association of passion with unfreedom. In the *congedo* to the *Canzone montanina*, Dante writes that he is so bound by the chain ("catena," *Amor, da che convien*, line 82) of love that were the cruelty of Florence to abate, and his banishment be lifted, he would still not be free to return.[52] In *Paradiso* 26, the "corde" of love strikingly invert the lyric formula: being drawn by the cords of love signifies not the unfreedom of passion but the concord of human and divine wills in charity ("Tutti quei morsi… a la mia caritate son concorsi," *Par.* 26.55–57). The association of cord, concord, and *caritas* has roots in sources as varied as Seneca, Cicero, Isidore, Augustine, Pseudo Hugh of St. Victor, Richard of St. Victor, and even medieval preachers contemporary to Dante (Ruini 2007). Dante is squarely within this tradition, since he holds that only the grace of charity quiets the human will, and the concord that results from *caritas* is, as he says in the *Monarchia*, nothing other than the unity of wills (*Par.* 3.70–71, 79–81; *Mon.* 1.15.8).[53]

When Dante responds to St. John, saying that the bites of love turn his heart to God ("cor volgere a Dio," *Par.* 26.56), the sense in which he uses the word "heart" is as the "secreto dentro" or the "secret within," not a particular faculty of the soul or body, but the "innermost center" (*Conv.* 2.6.2; Ralby 2021: 26; Gentili 2002). Insofar as the heart is turned to God, "cor" cannot retain the Aristotelian sense it has in Peter of Spain, Giacomo da Lentini, or as seen above in Cecco d'Ascoli: as a physical organ and/or seat of sense-perception, interfacing between soul and body (Gentili 2002; Pace 2020). Rather, here "cor" should be taken as Dante uses it in the *Convivio*, in its Augustinian sense as the soul's inmost interiority (Gentili 2002: 4; Ralby 2021: 18).[54] As we will see in the final image, the "caritate" (line 57) or divine love that lies within Dante's heart allows his will to rest in concord ("concorsi," line 57), in union, with the divine will (*Par.* 3.70–71; *Mon.* 1.15.8; Ralby 2021: 26–7).[55] Repurposed from his own lyrics, the motifs of the bites and cords of love show Dante is equally willing to render poetically the grace of charity as violent *ekstasis* and quiescent *concordia*, transfiguring the language of passion into that of divine love.

The lyric language of passion and compulsion is also effective for articulating the paradoxical sense of necessity and freedom at play in the union of human and divine wills in *caritas*. As Marco Lombardo says, "a miglior natura / liberi soggiacete" (*Purg.* 16.79–80), you free ones are subject to a higher nature; likewise, Piccarda says "s'essere in carità è qui *necesse*" (*Par.* 3.77), being in charity is necessary to being in paradise. This formulation of freedom and necessity has an Augustinian lineage. For Augustine, the complete freedom of the will is not an unlimited capacity for action, but a particular incapacity—*non posse peccare*—the inability to sin (Komline 2020: 385). As Dante himself says in the *Monarchia*, the immutability of the wills of both the angels and the *beati*, i.e. their inability to sin and turn away from the will of God, indicates not that they have lost their freedom but that they have perfected it (*Mon.* 1.12.5). The adherence of the will of the beatified soul to God's will is paradoxically a "voluntary and blessed necessity" (*uoluntaria felixque necessitas*) (Komline 2020: 395). In *Paradiso* 26, Dante shows how in the voluntary and blessed necessity of charity, the soul is bitten by love and freely drawn up to God in love's bonds.

Desire, Will, and Passionless Love in the Final Image

The most definitive illustration of passionless love in the *Commedia* is the final image of the *Paradiso*, where love moves the pilgrim's desire and will like the uniform rotation of a wheel:

> A l'alta fantasia qui mancò possa;
> ma già volgeva il mio disio e 'l *velle*,
> sì come rota ch'igualmente è mossa,
> l'amor che move il sole e l'altre stelle. (*Par.* 33.142–45)[56]

Figuring how divine love moves him in the same way it moves the stars, Dante's image of the wheel of his desire and will has garnered more than a century of critical debate.[57] That *disio* and *velle* are near synonyms has even led some interpreters to read the poem's ending as amounting to empty tautology (Ledda 2002: 319). Addressing the shades of meaning of *disio* and *velle* will clarify how the final image is far from tautological and will shed light on Dante's notion of intellective love. A dominant proposal in the critical accounts of the final image is Freccero's, which reads *disio* as the desire of the intellect, and *velle* as the will's "perfect act of fruition" (Freccero 1964: 26). Chiavacci Leonardi, however, reads both *disio* and *velle* as aspects of the appetitive faculty. Her counter-proposal is that *disio* stands for "desiderio o impulso naturale," desire or natural impulse, while *velle* means "volere deliberato" or deliberative will (Chiavacci Leonardi

2016: 3.929). In contrast to Freccero's referral of *disio* to the intellect, Chiavacci Leonardi stands on firmer philosophical ground by taking both *disio* and *velle* as aspects of appetition, and the analysis below will follow this part of her exposition. However, according to the Thomistic doctrine Chiavacci Leonardi relies upon, her gloss of *velle* as deliberative will does not hold up.[58]

"Volere deliberato" or *voluntas deliberativa* is the act of will, undertaken with discursive reasoning or judgment, that chooses between particular goods (Kent 1995: 231; *ST* 1.2, q.18, a.9). Chiavacci Leonardi equates "volere deliberato" with Aquinas' notion of *dilectio electiva*, the love of rationally chosen particular goods, ordered to an ultimate end (*ST*. 2.2, q.23, a.1). But neither reason nor particular goods apply to the final vision because the final vision regards the universal Good, God. *Caritas* "goes beyond the rule of human reason," writes Aquinas, and is, therefore, not expressed in the soul's faculty of rational choice, *liberum arbitrium*, but in the will, which regards the ultimate end itself, *ipsius finis* (*ST.* 2.2, q.23, a.1). The love of God is not the love of a particular good chosen in view of an ultimate end, but the love of that ultimate end itself, the *summum bonum*. Likewise, Dante could not be clearer: the "fantasia" (*Par.* 33.142), the apprehensive basis of discursive reasoning, loses all power in the vision of God. While "volere deliberato" does convey an "intellectual" dimension of will, it does so only with regard to reasoned choice about particular goods, that is, about everything that is *not* God.[59] For Dante, the love of God is not a matter of discursive deliberation but the non-discursive union of the soul to God.

If Chiavacci's reading of the final image breaks down with its intellectualized notion of *velle* as discursive, rational choice, Freccero's runs aground on his intellectualized notion of *disio* (Freccero 1964: 24). Freccero misconstrues *disio* as referring to the immanence of "personal fulfillment," of the intellect's desire to know God (Freccero 1964: 24). Likewise, Freccero misconstrues *velle* as signaling transcendence by mischaracterizing the will as a faculty always outwardly turned to its object. There are two main reasons why this account of *disio* and *velle* should be modified and reversed. The first is that Freccero's Neoplatonic framework misses the Aristotelian correspondence Dante establishes between the motion of his own desire and that of the celestial spheres: both are moved, according to Aristotle and Dante, by their *disio* for the Unmoved Mover. The second limitation is that Freccero's insistence on the physicality of the image of the wheel leads him to externalize the motion of *velle*, even as he recognizes the perfection of the will is in its immanent fruition and enjoyment of divine love, and even though throughout the *Paradiso* Dante figures *caritas* as the *inward* interpenetration of human and divine wills (Freccero 1964: 19–24). *Pace* Freccero, in the final image of the wheel, *disio* figures the transcendent and *velle* the immanent dimensions of the love of God. *Disio,* as the motion toward that which is loved, figures how God is both the transcendent aim of the soul's love and the transcendent love that moves or draws the soul to God. *Velle,* the seat of *caritas*, figures the immanence of God's presence to the soul and the soul's presence to God through divine love, which perfects or completes the motion of the soul's desire for God in the still point of union (*ST* 2.2, q.24, a.1).[60]

If we are to heed Freccero's own call for a more literal understanding of the final image, rather than stretching *disio* to mean the intellect's desire for knowledge, *disio* should be left to simply mean desire (Freccero 1964: 16–18). Read more closely to the letter, desire regards the good not yet possessed and is "simply the motion toward the object of desire" (*ST* 1.2, q.30, a.1). As Vergil himself says in his discourse on love, desire is a "spiritual motion" of the lover toward the beloved, which does not rest until the lover enjoys the attainment of that which is loved (*Purg*. 18.31–33; *ST* 1.2, q.26, a.2). Love is thus the principle of desire's movement; love draws, attracts, or bends the lover to the beloved ("quel piegare è amor," *Purg*. 18.26).[61] In this sense, the love of God is the transcendent principle of the motion of Dante's *disio.* Like a gravitational pull, divine love bends and draws Dante's desire in centrepetal motion to its transcendent end, God as love itself (Barsella 2010: 98–99). *Disio,* not *velle*, images the "outer" motion Freccero calls the "circular track" or orbit into which Dante's soul is drawn to God, in harmony with the angelic intelligences, celestial spheres, and the rest of creation (Freccero 1964: 26).

Dante's *disio* moves in harmony with the angelic intelligences and celestial spheres because they are moved by love in the same way. As Falzone points out, referring to Nardi, Dante's phrase indicating the uniform motion of the wheel, "igualmente è mossa," vernacularizes the various renderings of uniform celestial motion in the Latin Aristotelian tradition, such as *motus regularis, motus uniformis*, and *motus aequalis* (Falzone 2010b: 49). According to the Aristotelian doctrine that influences Dante, the Unmoved Mover moves the celestial spheres as something loved, "movet autem ut amatum" (*Met.* 12.10, 72b 3). In his profession of faith to St. Peter, Dante follows Aristotle virtually to the letter:

> [...] "Io credo in uno Dio
> solo ed etterno, che tutto'l ciel move,
> non moto, con amore e con disio." (*Par.* *24*.130–32)[62]

Because the Unmoved Mover is not itself in motion, it does not cause movement in the same way that things already in motion do. To move something as something loved means to move it by attraction, that is, by effecting *disio* or *appetition* in that which is moved (*ST* 1.2, q.26, a.3). Derived from the prefix "ad," meaning toward, and "petere" to aim at, reach for, or desire, appetition in its broadest sense signifies the "tendency for anything to seek what completes it"; in this sense appetition is a motion from potency to act (Miner 2009: 16). Everything except God—who is complete within himself because he is his own act of being—is composed of some degree of potentiality and actuality, and, therefore, seeks that which can fully actualize its being. Like the motion of the angels and the spheres, the motion of Dante's *disio* is toward what it loves, the Unmoved Mover, God, the *amatum* that is the principle of their being. As far as their nature allows, the desire of the soul and the angels is to become one with their ultimate principle. This is why, as the sphere nearest to the Unmoved Mover, the Primo Mobile is the "ciel velocissimo," the swiftest sphere. The Seraphim who move the Primo Mobile are the angels closest to and have the most intense desire for union with God (Barsella 2010: 62–65). The intensity of the Seraphim's desire for God corresponds to the velocity of the motion they impart to the Primo Mobile (*Par.* 27.99). Like the Primo Mobile, which is turned by the love of God, the love of God turns Dante's *disio* by being his desire's aim and end (*Par.* 27.111; *Par.* 33.143–45).

The motion of Dante's *velle*, according to Freccero, is not circular but has a "rectilinear thrust" (Freccero 1964: 18). His reading derives from the physicality he ascribes to Dante's wheel, as well as the Neoplatonic account of angelic motion found in Book 15 of Pseudo-Dionysius' *Celestial Hierarchy*. However, this physicality is operative neither in Dante's notion of the will nor in the Dionysian account of angelic motion.[63] Dionysius' angels, which are described as winged wheels (an elaboration of Ezekiel 1:18–21), indeed move "straight ahead," but not because they are physical wheels that have "tangency with the ground" as Freccero would have it (Freccero 1964: 18). After all, in Pseudo-Dionysius, angels are figured as winged wheels flying skyward in completely spiritual or immaterial motion. The Areopagite is unmistakably clear, as Freccero's own quotation demonstrates: the angels ceaselessly circulate "around the Good" and their motion directed to God is "not of this world" (*supermundane*) (Freccero 1964: 18). Thomas Gallus—whose commentaries on Pseudo-Dionysius were used by the likes of Albert the Great, Aquinas, and Bonaventure, and whose paraphrase of the entire Dionysian *corpus* (the *Extractio*) was included in the standard textbook at the University of Paris (Gallus 2022: 15–16, 24–25; Coolman 2017: 5–6)—glosses the linearity of angelic motion as a "direct and upright path" (*viam rectam et erectam*), an ascension or "upward-lifting extension into God," and as a "sharp penetration into the divine hiddenness" (Gallus 2022: 259).[64] Here Gallus stresses how angelic motion is *rectus* not because it is rectilinear, but because it is a direct, immersive movement of union into and within divinity, never turning away from the love and desire of God (*indeclinabilem a desiderio e amore Dei*) (Gallus 2011: 712; 2022: 259). Like Gallus, Dante understands *rectus* to indicate the "right aim of the will" (*directio voluntatis*), and as he says in the *De vulgari eloquentia*, he considers himself to be a poet of "rectitude" (*rectitudinem*) (*DVE* 2.2). For Dante, the right aim of the will is right love: having lost the straight way ("diritta via," *Inf*. 1.3) and fallen into the sea of twisted love ("del mar de l'amor torto," *Par.* 26.62), in paradise Dante arrives on shores of "l'amor diritto," right love (*Par.* 26.63). However, like the angels, in the final vision Dante's upright *velle* does not merely aim at the transcendent love of God, but immanently reaches and holds steadfast to it.

The *Amor* that turns the pilgrim's *velle* signifies the immanent presence of God to the soul, and the turning of the pilgrim's *velle* signifies the soul's presence to the love of God it receives. In the final vision Dante receives an infusion or flash of divine grace ("un fulgore") that fulfills his will/desire ("voglia," *Par.* 33.141).[65] This infusion or flash of grace begets the interior act of charity that unites the human will to God (*ST* 2.2, q.27, a.7). Because the will, and not the sensitive appetite, is the seat of charity, *velle* can thus be read as the faculty of the soul that, infused with grace, brings Dante's *disio* to its perfection and rest in divine love (*ST* 2.2, q.24, a.1). Since the motion of desire orients toward the good not yet possessed, *disio* signifies the soul's longing for God as the transcendent, ultimate Good. *Velle* conveys the sense in which this good is attained. As Falzone puts it, *velle* expresses "the concrete adhesion of the will to that which perfects the *disio*."[66] The act of the will is, in fact, nothing other than "actualis appetitus boni," the desire of the good made actual (*ST* 1.2, q.8, a.1; Nardi 1944: 337–8).

Elsewhere in the *Commedia* Dante uses "volontà" as the vernacular form of *voluntas*, the Latin noun for the will, but in the final verses Dante uses "*velle*," the Latin infinitive meaning "to wish or will."[67] As a nominalized verb, *velle* captures how for Dante the will is a faculty (noun) that is one with its operation (verb): the actualization of desire. Moreover, the fact that Dante uses the possessive pronoun for desire ("il mio disio") but the impersonal article for will ("il *velle*," *Par.* 33.143) expresses how in the *visio Dei*, Dante's and God's will are but one will. In fact, for Dante the very idea of uniform motion connotes the unity of wills ("unitatem voluntatum, que per uniformem motum datur intelligi," *Mon.* 1.15.5). Dante's *disio* or longing for God is properly his own because it is of his human nature: *disio* is an expression of the potency of his created being moving toward the perfection of its act, that is, adhering to the being and love of the Creator. *Velle* signals the actualization of this *disio*, how the pilgrim's desire is completed and brought to rest in the experience of his own being as one with its ground, one in the love of God and union of will.

The wheel of desire and will is thus a paradox of motion and stillness. The movement of *disio* for the love of God is simultaneously brought to rest by the love of God infused in the *velle*. Augustine captures this paradoxical simultaneity of stillness and motion, longing and fulfillment, when he describes the love of God as *insatiabili satietate*, a "desire always satisfied but never sated" (*Confessions* 2.10.18). In rendering the stillness-in-motion of *disio* and *velle*, Dante is also in accord with Pseudo-Dionysius: the angel's circulation around the Good is a "motu semper immobili," a motion ever unmoving (Gallus 2022: 260). The love that moves Dante's desire also brings rest and fulfillment to the will. The grace of charity, the "virtù di carità" Piccarda says, "quieta la nostra volontà," quiets and fulfills our will (*Par.* 3.70–71). Through *caritas*, the will reaches what it wills, without desire or thirst for anything else.[68] The love of God permeating the will perfects and brings desire to rest.

The final image can be described thus: in the vision of God, the soul is caught in the circular orbit of God's love, the impetus and terminus of the soul's movement, attracting the soul through its *disio* and completing it through its *velle*. There is a likeness here between the unitive circularity of intellective love and the circularity of passion. In passion the "appetible object moves the appetite, introducing itself, as it were, into its intention... the appetite moves towards the realization of the appetible object, so that the movement ends where it began" (*ST* 1.2, q.26, a.2). Similarly, the wheel of desire and will figures a circle of immanence and transcendence: the love that moves Dante's *disio* and *velle* is the very love to which they are moved. But there is a crucial difference between love that is a passion and love that is divine: the passion of love is mediated and never reaches God, while divine love is unmediated and joins the soul to God. In passion, the sensible object stirs the appetite of the subject, but indirectly through the medium of perception. By contrast, in passionless love, made possible through the soul's *sanza mezzo* creation, there is immediate contact between the soul and God.[69]

The harmonious motion of desire and will in the final image also illustrates the apparent paradox that Dante could never have experienced passionless love without his passionate love for Beatrice, and without, in some sense, moving beyond it. Knowing Dante better than himself, Beatrice sends St. Bernard to the pilgrim as his final guide, so that his desire may be fulfilled: "A terminar lo tuo disiro / mosse Beatrice me del loco mio," says Bernard (*Par.* 31.65–66). So that Dante's desire reaches its end, Beatrice removes herself as the object of the pilgrim's desire. While the driving force of both the pilgrim's spiritual journey and the poet's poem, Dante's desire for Beatrice is preparatory for the final vision. As Beatrice herself says, "paradise is not in my eyes alone."[70] By sending Bernard and removing herself as the object of his love, Beatrice enacts in Dante a transformation from one love to another that is parallel to *Epistola* 3, but of a higher ontological order: not from one passion to another, but from love of the created to love of the Creator, from herself to God (*Purg.* 31.49–57). Though she is not subject to corruption in her blessed state, Beatrice is nonetheless a *cosa creata*; she cannot be that which fully actualizes Dante's *disio* and *velle*. Only God can "terminar" or perfect Dante's "disiro" (*Par.* 31.65).

However, in the love of God, Dante's love for Beatrice is not erased or corrupted, as it was in *Epistola* 3. As Cunizza da Romano and Folquet de Marselha attest in the heaven of Venus, all love, including erotic passion, is recuperated in paradise (*Par.* 9.34–36, 103–5); so too is Dante's love for Beatrice. Before the final vision, Dante expresses a helplessness at his inability to express his love for Beatrice, that even if he distilled everything he had written of her into one word of praise, it would fail to capture her beauty because, surpassing human nature, it can be fully enjoyed only by its "fattor," its maker.[71] From the passion that sustains his lyrics, tohis love for her that even "imparadises" Dante's mind in the *Commedia*: both are incommensurable to Beatrice (*Par.* 28.3). Only the Creator can fully love his creation.

Yet, the nature of divine love, *caritas*, is that it "begins with God as the last end, and passes on to creatures" (*ST* 2.2, q.27, a.4). But how is this possible? As Piccarda says, each among the blessed is enveloped in the divine will, and in union with God, all wills are made one (*Par.* 3.80–81).[72] In other words, the union of will between each blessed soul and God results in the unity of wills among the blessed. St. Bernard's prayer to the Virgin illustrates this: Bernard says he burns for Dante's vision of God more than he did for his own, as Beatrice and the other *beati* join him in prayer on Dante's behalf.[73] This unity is seen also when, with remarkable poetry, Dante neologizes pronouns into verbs to convey his longing for the mutual indwelling of his being with that of Folchetto: "s'io m'intuassi, come tu t'inmii," if I were to in-you myself, as you in-me yourself (*Par.* 9.81). Once Dante "in-Hims" ("inluia") his own will with God's, he may share that same interpenetration of being with the other *beati*, including Beatrice (*Par.* 9.73).

Even when he experiences the profound separation of Beatrice departing as his guide, the reality of being in the Empyrean, which is beyond the limitation of space and time, allows Dante to experience the unity underlying separation and difference. Dante describes the distance between him and Beatrice as greater than what any "mortal eye" could experience, greater than the distance that spans the depth of the ocean to the high point of the sky (*Par.* 31.73–75). Yet this distance ultimately amounts to nothing ("ma nulla mi facea," *Par.* 31.77), since Beatrice still appears to Dante in a direct, unmediated way ("sua effige / non discendea a me per mezzo mista," lines 77–78). This unmediated relationship is possible because in the Empyrean, which does not admit of determinate position let alone separation in time or space, the blessed souls dwell united to each other *through* their union with God in the ground of being (Moevs 2005: 28).

In other words, the love that unites Dante's will to God's is the same love that unites Dante to Beatrice. Love of the Creator is transitive to the love of creation, as Dante's final guide puts it, in the relationship between self and neighbor: "You must first love God, so that in him you can love your neighbor too" (Bernard of Clairvaux 1987: 193). The testament of the *Commedia*, however, reverses Bernard: it is through the love of the neighbor, his amorous passion for Beatrice and her passionless love for him, that Dante arrives at the love of God.

Notes

1. Derived from Aristotle's *Gen.* 1.7, this *sententia* was widespread in the scholastic discourse and *florilegia* that circulated in Dante's time, like the Latin compendium *Auctoritates Aristoteles* (Fioravanti 2019). Dante could have also heard this quotation used at a public quodlibetal debate held by Peter of Trabibus, the *lector* at Santa Croce during the mid-1290s (Dell'Oso 2022).
2. It should be noted, however, *pace* Foster and Boyde that the *Vita nuova* also places great emphasis on the effects of love on the body, especially in the opening of the *libello*, which draws on the pneumatic doctrine of the spirits deriving from Galen and mediated by Albertus Magnus.
3. *Conv.* 3.3.1: "dare a intendere quale amore è questo, per lo loco nel quale adopera." English translations adapted from Frisardi (2018).
4. On the soul's self-awareness Dante echoes the widespread scholastic teaching that only the soul's intellectual powers have awareness of their own acts. Reason, will, and free choice (*liberum arbitrium*) are self-reflexive because they operate independently of the body. See Thomas Aquinas: "Nulla autem potentia iudicans seipsam ad iudicandum movet nisi supra actum suum reflectatur… Quod quidem solius intellectus est [But no judging power moves itself to judge unless it reflects on its own action… and this only the intellect can do]" (*SCG* 2.48.3). And Bonaventure: "Si igitur omnes potentiae sunt alligatae materiae et substantiae corporali praeter solam rationem, sola illa est, quae potest se super se ipsam reflectere [If, therefore, all faculties are bound to matter and corporeal substance except reason alone, it (reason) alone is able to reflect on itself]" (2 *Sentences* d. 25, p.1, a.1, q.1). English translations of the *Summa contra gentiles* are from Anderson (1975). English translations of Bonaventure's *Sentences* are my own.
5. "Circa alias vero animae operationes, sicut est amare, gaudere, et alia huiusmodi, est aequivocatio cavenda. Nam quandoque sumuntur ut sunt animae passiones. Et sic sunt actus sensibilis appetitus secundum concupiscibilem vel irascibilem, cum aliqua

permutatione corporali… Sumuntur autem quandoque pro simplici actu voluntatis, qui est absque passione… Cum vero voluntas sit potentia non utens organo, sicut nec intellectus, palam est huiusmodi, secundum quod sunt actus voluntatis, in anima separata remanere" (*SCG* 2.81.15).

6. To refer to affects of the will, including the intellectual love of God, Peter King uses the infelicitous term "pseudopassions": "these pseudopassions, unlike the passions, do not involve any somatic reactions or indeed any material basis at all. They are located in the intellective appetite as rational acts of will. Angels and disembodied human souls experience only these pseudopassions; animals experience only passions; living human beings alone are capable of both. The *amor intellectualis Dei* is a pseudopassion, one that may be deeply held" (2008: 105). Nicholas Lombardo uses the phrase "affections of the will" (2011: 76–77). For an informative and deeply interesting discussion of Aquinas' treatment of the difference between the passions and the spiritual affects that accompany the Gifts of the Holy Spirit, see Stump (2011: 39–43). The question of the will as a locus of affect has its roots in Augustine's account of the passions in *The City of God* 9 and 14. In fact, Augustine considers passions themselves to be *voluntates* or instances of willing: "Voluntas est quippe in omnibus, immo omnes nihil aliud quam voluntates sunt. Nam… in hos vel illos affectus mutatur et vertitur voluntas [the will is therefore in all of them, and all the passions are nothing other than willings. For… will shifts and turns into these or those affections]" (*Civ. Dei* 14.5).
7. These are the four primary passions formulated by the Stoics: *cupiditas*, *timor*, *laetitia*, *tristitia* (desire, fear, joy, and sorrow). They were widely adopted in the Middle Ages, notably by Boethius in the *Consolation of Philosophy* (1.7) and Augustine in the *City of God* Books 9 and 14. On the Stoic account of the passions see Graver (2007). On medieval theories of emotion see Knuuttila (2004).
8. In his commentary on the *Viaticum*, the authoritative text on lovesickness (*amor heroes*) in the medical curricula at European universities in the twelfth and thirteenth centuries, Peter of Spain specifies the relationship between passion, accident, and the pathology of love in the following way: "omnia accidencia anime sunt passio cordis… Sed amor eroes est accidens. Ergo est passio

cordis [all accidents of the soul are passions of the heart...But lovesickness is an accident. Therefore it is a passion of the heart]" (quoted in Tonelli 2015: 41). Like Peter, Cavalcanti treats love as a psycho-physical phenomenon explained in terms of Aristotelian natural philosophy. *Donna me prega*'s "natural dimonstramento" of love defines love as an "accidente" (1–2), which Dino del Garbo—the physician who wrote what is likely the first commentary on the canzone—glosses as "passione": "Deinde subdit 'd'uno accidente' etc. ubi hac passione de qua tractare intendit... [then he continues 'of an accident,' about which passion he intends to treat...]" (Fenzi 1999: 89).

9. For engagements with Trabibus' philosophy and theology see especially Simoncioli (1956), Di Noto (1963), Huning (1968), Houser (1979), Doyle (1984), Piron (2008), Friedman (2019, 2020), Friedman (forthcoming), and Vaura (2022).
10. *Ep.* 3.2: "that is to say, from one passion to another, the nature of the passion remaining the same, but the objects differing in number, not in kind." English translations of *Epistola* 3 are adapted from Toynbee (1966).
11. Baglio glosses the "objects" of the concupiscible power as "più donne," other women, even though the concupiscible power, strictly speaking, is not limited to romantic objects of love; it regards any kind of sense-based good (2016: 83).
12. See below for how *appetitus sequitur apprehensionem* is dramatized in Vergil's discourse on love in *Purg.* 18.
13. See Dante in *Mon.* 1.12.3: "nam primo res apprehenditur, deinde apprehensa bona vel mala iudicatur, et ultimo iudicans prosequitur sive fugit [for first a thing is perceived, then it is judged to be good or evil, and finally the person who judges pursues it or shuns it]." English translations from Shaw (1996).
14. *Ep.* 3.5: "For, every faculty which is not destroyed after the consummation of one act is naturally reserved for another. Consequently, the faculties of sense, if the organ survives, are not destroyed by the consummation of one act, but are naturally reserved for another."
15. *SCG* 2.55.4: "In every instance of corruption, furthermore, potentiality remains after the removal of act."
16. See Aristotle's definition of alteration: "alteratio quidem est, quando, manente subiecto sensibili existente, transmutatur in eius

passionibus, aut contrariis entibus aut mediis: puta corpus sanum est, et rursus laborat, manens idem [here is 'alteration' when the substratum is perceptible and persists, but changes in its own properties, the properties in question being opposed to one another either as contraries or as intermediates. The body, e.g. although persisting as the same body, is now healthy and now ill.]" (L. 10.73, *De gen.* 319b 4).

17. This question is found in Assisi, Biblioteca Comunale MS 154, fol. 24va-28ra, which contains the *Ordinatio or revised version* of Book One of Peter's *Sentences Commentary*. See Friedman (forthcoming) for discussion of the significant overlaps, particularly on this question of the plurality of forms, between this manuscript and Florence, Biblioteca Nazionale Centrale, Conventi Soppressi D. VI. 359, the manuscript containing Trabibus' work that is most directly connected to Dante's Florence. One way to put the scholastic debate over the plurality or unicity of forms is as follows. Does the human being have a single substantial form or multiple? The unicity position holds that the soul is a unitary form that integrates all the human being's faculties and enlivens the body. The plurality position holds that the human being is a composite of multiple forms, which give a substantial consistency to the body apart from the soul, and which allow the soul's faculties to operate on differing planes of existence and experience, e.g. that of the vital, the sensible, and the intelligible.
18. *Conv.* 4.10.9: "everything that is corrupted, does so by way of a preceding alteration."
19. *Ep.* 3.4: "amorem huius posse torpescere atque denique interire, nec non huius, quod corruptio unius generatio sit alterius, in anima reformari."
20. *Acerba* 3.1.49–51: "Senza vedere l'uom pò inamorare / formando specchio di la nuda mente, / vedendo vista su nel'maginare [without seeing (her) a man can fall in love, forming a mirror in the naked mind / seeing her image in the imagination]." My translations.
21. *Io sono stato*, lines 1–2: "I have been together with love since my ninth circling of the sun." English translations of Dante's lyrics are from Foster and Boyde (1967).

22. *Io sono stato*, lines 12–14: "Indeed Love can prick the flank with new spurs; and whatever is the pleasure that leads us now, it is necessary to follow, if the other is outworn."
23. *Purg*. 18.22–23: "your faculty of apprehension draws an intention from a real being." English translations of the *Commedia* are adapted from Mandelbaum (1982–1986).
24. *Purg*. 18.23–24: "e dentro a voi la spiega, / sì che l'animo ad essa volger face."
25. *Mon*. 1.12.3: "Et ideo dico quod iudicium medium est apprehensionis et appetitus."
26. For the importance of this canzone to Dante's engagement with the lyric tradition, the ethics of love he elaborates in *Inferno* 5, and his divergence from Cavalcanti, see Barolini (2006: 70–101).
27. *Io sono stato*, lines 10–11: "liber arbitrio già mai non fu franco, / sì che consiglio invan vi si balestra."
28. Similarly, *Conv*. 4.12.5 likens the soul's desire to a pilgrim who, while traveling on an unknown road, moves from house to house, believing each to be an inn where he can find rest; so the soul moves from earthly good to earthly good until it rests in the inn of the *sommo bene*.
29. Here I extend to *Amor che nella mente* the distinction Barolini makes between Dante's poetics in the *Vita nuova* and Guinizelli's (2006: 32).
30. *Conv*. 3.14.4: "è da sapere che lo primo agente, cioè Dio, pinge la sua virtù in cose per modo di diritto raggio... onde ne le Intelligenze raggia la divina luce sanza mezzo [it should be understood that the first agent, God, impels his power into some things by means of a direct ray... Thus the divine light radiates into the separate Intelligences without any intermediary]."
31. *Conv*. 3.6.6: "ch'ella è così fatta come l'essemplo intenzionale che de la umana essenzia è ne la divina mente."
32. *Conv*. 3.6.6: "la forma umana, in quanto ella è per intenzione regolata ne la divina mente... è perfettissima ne la umana generazione."
33. *Conv*. 3.6.6: "E se essa umana forma, essemplata e individuate, non è perfetta, non è manco dello detto essemplo, ma della materia la quale individua."
34. The fact that the sun shines on her and that she speaks "here below" demonstrates the *donna* of the canzone to be flesh and blood, not a purely ideal object: "Non vede il sol, che tutto'l

mondo gira, / cosa tanto gentil, quanto in quell'ora / che luce ne la parte ove dimora / la donna di cui dire Amor mi face... Quivi dov'ella parla..." (lines 19–22, 41).

35. *Par.* 7.124–29: "You say: 'I see water, I see fire and air and earth, and all their mixtures come to corruption and last but a little while, and yet these too were created things; so that, if what has been said is true, they ought to be secure against corruption'."
36. *Par.* 7.67–72: "All that derives directly from this Goodness is everlasting, since the seal of Goodness impresses an imprint that never moves. Whatever rains from it immediately is fully free, for it is not subject to the power of new things."
37. *Par.* 7.142–44: "But your life is breathed forth immediately by the Chief Good, who so enamors it of His own Self that it desires Him always."
38. I am grateful to Russell Friedman for sharing forthcoming work on Peter of Trabibus, which not only synthesizes the existing scholarship, but also adds convincing new evidence to the dating and identification of the manuscripts containing Trabibus' extant *ouevre*, especially the ones most closely connected to Dante's Florentine milieu: Trabibus' *Quodlibetales and Lectura of his Sentences commentary* found in Florence, Biblioteca Nazionale Centrale, MS Conventi Soppressi D. VI. 359, as well as the *Ordinatio* or revised *Sentences* commentary found in Florence, Biblioteca Nazionale Centrale, MS Conventi Soppressi B.V.1149. See also Piron (2008: 75–86) and Huning (1964: 209–11, 221).
39. Florence, Biblioteca Nazionale Centrale, MS Conv. Soppr. B.V.1149, fol. 128ra: "quod impossibile est liberum arbitrium cogi sive ab agente creato sive increato."
40. *Purg.* 21.43–60: "This place is free from all alteration: only that which heaven receives from itself and in itself can serve as cause here, and nothing else... it trembles here when some soul feels it is cleansed, so that it rises or stirs to climb on high; and that shout follows."
41. See both books of Aristotle's *On Generation and Corruption* for references to the passions of the elements and their alteration. For example, "Passiones enim, secundum quas hoc dicimus contingere, differentiae elementorum sunt: dico autem puta calidum et frigidum, album et nigrum [For alteration, as we assert, takes place in respect to certain qualities: and these qualities (I mean, e.g.

hot–cold, white-black, dry-moist, soft-hard, and so forth) are, all of them, differences characterizing the 'elements']" (L. 2, *De gen.* 314b 18–21).

42. *Purg*. 21.61–63: "Of its purity only the will gives proof, completely free to change its dwelling, the soul surprises itself, and rejoices in its will."
43. *Purg*. 21.64–66: "First the soul wills the good, but the desire that divine justice puts to the punishment, against the will, to the degree it was set to sin, does not permit the soul to ascend."
44. As shown above, for Dante the soul's vegetative and sensitive powers are united to the intellective powers, which God fashions as "un'alma sola" (*Purg*. 25.74). Trabibus holds the same idea, that the soul's powers are rooted in one and the same essence: "appetitus sensitivus movet, scilicet, inclinando voluntataem ad suum actum, quia cum isti appetitus sint radicati in una essentia animae, quando trahitur unus, per colligantiam trahitur alius, qui quidem tractus est magna dispositio ad inclinandam voluntatem ad volendum [the sensitive appetite moves the will, that is, by inclining the will to its act; this is because, since these appetites are rooted in one essence of the soul, when one is drawn the other is drawn by their being bound together, which act of being drawn is indeed a great disposition inclining the will to willing]" (Simoncioli 1956: 232).
45. English translations of the Summa *theologiae* are from the Blackfriars edition (Aquinas 2006).
46. I am grateful to Rohini Ralby for this point. See also Dante's terzina-length *ars poetica* in *Purg*. 24.52–54: "Io mi son un che, quando / Amor mi spira, noto, e a quel modo / ch'e' ditta dentro vo significando [I am one who when love breathes within me, I take note, and in the way it is dictated within, I go signifying]."
47. In *Paradiso* 29 Dante says outright that affect follows intellect: "Onde, però che a l'atto che concepe / segue l'affetto, d'amar la dolcezza" (*Par.* 29.139–40).
48. *Par.* 26.49–51: "But say still more, if you feel other cords that draw you to him, so that you can proclaim how many teeth there are with which this love bites you."
49. *Par.* 26.55–57: "All those things whose bite can turn the heart toward God concord in my love."

50. *Così nel mio parlar*, lines 31–32: "The death that chews on every sense with the teeth of love."
51. I owe this articulation to Rohini Ralby.
52. *Amor, da che convien*, lines 82–84: "Là ond'io vegno una catena il serra / tal, che se piega vostra crudeltate, / non ha di ritornare qui libertate [he's bound by such a chain in the place I come from that even were your harshness to relax, he is not free to return here]."
53. The definition of *concordia* in *Monarchia* 1.15.8: "omnis concordia dependet ab unitate, que est in voluntatibus."
54. Compare Dante's spiritual sense of the word "heart" with Cecco's material sense of the word, as seen above.
55. In addition to *Living the Practice,* I am indebted to many conversations with Rohini Ralby for this understanding of the relationship between the heart and will. On the relationship of *cor* and *voluntas in Augustine* see (Komline 2020: 254). See also Aquinas on the will as the subject of *caritas*: "Now charity's object is not some good of the senses, but divine good, known by the intellect alone. Consequently, charity is to be found not in the sense appetite, but the intellective appetite, that is to say in the will" (*ST* 2.2, q.24, a.1).
56. *Par.* 33.142–45: "Here my highest imagining failed; but already my desire and will were turned, like a wheel being moved evenly, by the Love that moves the sun and other stars."
57. See Hollander's commentary for summary and bibliography (2007), Dartmouth Dante Project, http://dantelab.dartmouth.edu. Accessed 15 February 2023.
58. See also, Pertile (2005: 267–68), who follows this reading of *disio* and *velle*.
59. Discursive reasoning is not used in angelic choice but in *human* choice, on which see again the texts from Aquinas' *Summa Theologiae* that Chiavacci Leonardi cites (*ST* 1, q.60, a.2).
60. I am greatly indebted to discussions with Rohini Ralby and David Soud for this articulation, and what follows, about the immanence and transcendence of the love of God in relation to Dante's image of the wheel of *disio* and *velle*.
61. See Porro (2014) and Falzone (2003). See also Aquinas on love as the moving principle of desire: "amor dicitur illud quod est principium motus tendentis in finem amatum ['love' denotes that

which is principle of motion towards the end loved]" (*ST* 1.2, q.26, a.1). Translation modified.

62. *Par.* 24.130–32: "I believe in one God, sole and eternal, who moves all the heavens, unmoved, with love and desire."
63. See Susanna Barsella's seminal study of Dante's angelology (2010), and, specifically, the influence of the Pseudo-Dionysian tradition upon it (2010: 27–69).
64. For the most comprehensive treatment of Gallus' theology see Coolman (2017).
65. Most medieval commentators agree that the *fulgore* that strikes the *mente* of Dante is divine revelation. The *Chiose ambrosiane* glosses *fulgore* specifically as divine grace: "Fulgore—Divine gratie illustrantis ad talem et tantam visionem," Dartmouth Dante Project, http://dantelab.dartmouth.edu. Accessed 15 February 2023.
66. Falzone (2010b: 58): "la concreta adesione del volere a ciò che perfeziona il disio." My translations. Freccero himself acknowledges "the unshakable adherence of the will," to that it loves (1964: 26).
67. For the vernacular forms of *volontà*, see for instance *Purg*. 21.69; *Par.* 3.70; *Par.* 4.76; *Par.* 5.22; *Par.* 15.68; *Par.* 19.86; *Par.* 32.63.
68. *Par.* 3.70–71: "fa volerne / sol quel ch'avemo, e d'altro non ci asseta."
69. See Aquinas, *ST* 2.2, q.27, a.4, ad. 3: "Et ideo caritas est quae, diligendo, animam immediate Deo coniungit spiritualis vinculo unionis [Consequently, charity, loving God, knits the soul to him without any intermediary in a bond of spiritual union]." Note again the use of the image of the *vinculo* to figure the union of the human and divine.
70. *Par.* 18.21: "non pur ne' miei occhi è paradiso."
71. *Par.* 30.16–21: "Se quanto infino a qui di lei si dice / fosse conchiuso tutto in una loda, / poca sarebbe a fornir questa vice. / La bellezza ch'io vidi si trasmoda / non pur di là da noi, ma certo io credo / che solo il suo fattor tutta la goda."
72. *Par.* 3.80–81: "dentro a la divina voglia, / per ch'una fansi nostre voglie stesse."
73. *Par.* 33.38–39: "vedi Beatrice con quanti beati / per li miei prieghi ti chiudon le mani [see Beatrice, with how many saints who fold their hands and join my prayers for you!]."

References

Alighieri, Dante. 1966. *The Letters of Dante*, ed. Paget Toynbee. Oxford: Clarendon.

Aligheri, Dante. 1966–1967. *La Commedia secondo l'antica vulgata*, ed. Giorgio Petrocchi, 4 vols. Milan: Mondadori.

Alighieri, Dante. 1980. *Vita Nuova*, ed. Domenico De Robertis. Milan: Ricciardi.

Alighieri, Dante. 1982–1986. *The Divine Comedy*, trans. Allen Mandelbaum. New York: Bantam Books. https://digitaldante.columbia.edu/. Accessed 10 June 2024.

Alighieri, Dante. 1996. *De vulgari eloquentia*, trans. Steven Botterill. Cambridge: Cambridge University Press.

Alighieri, Dante. 1996. *Monarchy*, trans. Prue Shaw. New York: Cambridge University Press.

Alighieri, Dante. 2005. *Rime*, ed. Domenico De Robertis. Florence: Edizioni del Galluzzo.

Alighieri, Dante. 2013. *Monarchia*. In *Le opere*, vol. 4, ed. Paolo Chiesa and Andrea Tabarroni. Rome: Salerno.

Alighieri, Dante. 2014. *Convivio*. In *Opere*, dir. Marco Santagata, vol. 2, ed. Gianfranco Fioravanti, 3–805. Milan: Mondadori.

Alighieri, Dante. 2014. *Dante's Lyric Poetry: Poems of Youth and of the "Vita Nuova" (1283–1292)*, ed. Teodolinda Barolini, trans. Richard Lansing. Toronto: University of Toronto Press.

Alighieri, Dante. 2016. *La Divina Commedia*, ed. Anna Maria Chiavacci Leonardi, 3 vols. Milan: Mondadori.

Alighieri, Dante. 2016. *Epistole, Ecloge, Questio de aqua et terra*. In *Opere di Dante*, ed. Marco Baglio, Luca Azzetta, Marco Petoletti, and Michele Rinaldi, vol. 5. Rome: Salerno Editrice.

Alighieri, Dante. 2018. *Convivio*, trans. Andrew Frisardi. Cambridge: Cambridge University Press.

Aquinas, Saint Thomas. 1961. *Summa Contra Gentiles*. Edited by C. Pera. Turin: Marietti.

Aquinas, Saint Thomas. 1964. *Exposition of Aristotle's Treatise on Generation and Corruption, Book I, cc. 1–5*, trans. Pierre Conway and Fabian R. Larcher. Columbus, OH: College of St. Mary of the Springs.

Aquinas, Saint Thomas. 1975. *Summa contra gentiles, Book 2: Creation*, trans. James F. Anderson. Notre Dame, IN: University of Notre Dame Press.

Aquinas, Saint Thomas. 2006. *Summa Theologiae*, 61 vols. New York: Cambridge University Press.

Ariani, Marco. 2009. Mistica degli affetti e intelletto d'amore: Per una ridefinizione del canto XXIV del Paradiso. *Rivista di studi danteschi* 9: 29–56.

Ardizzone, Maria Luisa. 2011. *Dante, il paradigma intellettuale. Un'"inventio" degli anni fiorentini*. Florence: Olschki.

Augustine. 1955. *De ciuitate dei*, ed. Bernhard Dombart and Alfons Kalb. Turnhout: Brepols.

Augustine. 2013. *The City of God: Books XI-XXII*, ed. Boniface Ramsey, trans. William S. Babcock. Hyde Park, NY: New City Press.

Augustine. 2014. *Confessions: Books 1–8, vol. 1*, trans. Carolyn J.-B. Hammond. Cambridge, MA: Harvard University Press.

Barolini, Teodolinda. 1992. *The Undivine Comedy: Detheologizing Dante*. Princeton: Princeton University Press.

Barolini, Teodolinda. 2006. *Dante and the Origins of Italian Literary Culture*. New York: Fordham University Press.

Barolini, Teodolinda. 2014. *Purgatorio* 21: De Amicitia. In *Commento Baroliniano*, Digital Dante. New York, NY: Columbia University Libraries.

Barolini, Teodolinda. 2022. *Dante's Multitudes: History, Philosophy, Method*. Notre Dame: Notre Dame Press.

Barsella, Susanna. 2010. *In the Light of the Angels: Angelology and Cosmology in Dante's Divina Commedia*. Florence: Leo S. Olschki.

Bemrose, Stephen. 1983. *Dante's Angelic Intelligences*. Rome: Edizioni di Storia e Letteratura.

Bernard of Clairvaux. 1987. *Selected Works*, trans. G.R. Evans. New York: Paulist Press.

Bonaventure. 1882–1902. *Opera Omnia*, vols. I-IV: *Commentaria in quatuor libros sententiarum Magistri Petri Lombardi*, eds. Fathers of the Collegium S. Bonaventurae. Ad Claras Aquas (Quaracchi): Ex typographia Colegii S. Bonaventurae.

Borzi, Italo. 1985. L'analisi dell'anima: amore e libertà (*Purg*. XVII). In *Verso l'ultima salute. Saggi danteschi*, 139–76. Milan: Rusconi.

Boyde, Patrick. 1993. *Perception and Passion in Dante's Comedy*. Cambridge: Cambridge University Press.

Byers, Sarah. 2012. *Perception, Sensibility, and Moral Motivation in Augustine: A Stoic-Platonic Synthesis*. Cambridge: Cambridge University Press.

Calenda, Corrado. 1995. *Potentia concupiscibilis, sedes amoris*: Il dibattito Dante-Cino. In *Appartenenze metriche ed esegesi: Dante, Cavalcanti, Guittone*, 111–24. Naples: Bibliopolis.

Cavalcanti, Guido. 1993. *Rime*, ed. Letterio Cassata. Anzio: De Rubeis.

Cecco d'Ascoli [Francesco Stabili]. 2016. *L'Acerba (Acerba etas)*, ed. Marco Albertazzi. Trento: La Finestra.

Cicero, Marcus Tullius. 1971. *Tusculan Disputations*, trans. J. E. King. Cambridge, MA: Harvard University Press.

Coolman, Boyd Taylor. 2017. *Eternally Spiraling into God: Knowledge, Love, and Ecstasy in the Theology of Thomas Gallus*. Oxford: Oxford University Press.

Corti, Maria. 1983. *La felicità mentale*. Turin: Einaudi.

Cullen, Christopher. 2006. *Bonaventure*. New York: Oxford University Press.

Dales, Richard. 1995. *The Problem of the Rational Soul in the Thirteenth Century.* Leiden: Brill.

Dante's Lyric Poetry. 1967. Trans. Kenelm Foster and Patrick Boyde, 2 vols. Oxford: Oxford University Press.

Davis, Charles T. 1965. Education in Dante's Florence. *Speculum* 40 (3): 415–35.

Davis, Charles T. 1988. The Florentine Studia and Dante's Library. In *The Divine Comedy and the Encyclopaedia of Arts and Sciences*, ed. Giuseppe Di Scipio and Aldo Scaglione, 339–66. Amsterdam: Benjamins.

Dell'Oso, Lorenzo. 2022. Dante, Peter of Trabibus, and the 'Schools of the Religious Orders' in Florence. *Italian Studies* 77 (3): 211–29.

Delorme, Ferdinand P. 1924. Pierre de Trabibus et la distinction formelle. *La France Franciscaine* 7: 255–69.

Di Noto, Antonio. 1963. *La théologie naturelle de Pierre de Trabibus: choix de questions du Ier livre des sentences Ms 154 de la Bibliothèque communale d'Assise.* Padova, CEDAM.

Dionysius the Areopagite. 1987. *Pseudo-Dionysius: The Complete Works*, trans. Colm Luibheid and Paul Rorem. New York: Paulist Press.

Doolan, Gregory T. 2008. *Aquinas on the Divine Ideas as Exemplar Causes.* Washington, DC: Catholic University of America Press.

Doyle, Patrick James. 1984. *The Disintegration of Divine Illumination Theory in the Franciscan School: 1285–1300: Peter of Trabes, Richard of Middleton, William of Ware.* Ph.D. Dissertation. Marquette University.

Elsky, Martin. 2001. Introduction to Erich Auerbach, "*Passio as Passion*" ["Passio als Leidenschaft"]. *Criticism* 43 (3): 285–308.

Falzone, Paolo. 2003. Psicologia dell'atto umano in Dante. Problemi di lessico e dottrina. In *Filosofia in volgare nel medioevo*, ed. Nadia Bray and Loris Sturlese, 331–66. Turnhout: Brepols.

Falzone, Paolo. 2010a. *Desiderio della scienza e desiderio di Dio nel Convivio di Dante.* Bologna: il Mulino.

Falzone, Paolo. 2010b. Visione beatifica e circolazione celeste negli ultimi versi del Paradiso. *Bollettino di italianistica* 7: 46–77.

Fenzi, Enrico. 1999. *La canzone d'amore di Guido Cavalcanti e i suoi antichi commenti.* Genova: il melangolo.

Fioravanti, Gianfranco. 2019. Come dice il filosofo: Dante e la '*Littera*' di Aristotele. *Italianistica* 48: 11–50.

Freccero, John. 1964. The Final Image: Paradiso XXXIII, 144. *MLN* 79: 14–27.

Friedman, Russell. 2019. Peter of Trabibus (fl. 1295), O.F.M., on the Physical and Mental Abilities of Children in Paradise. *Syzetesis* 6 (2): 433–60.

Friedman, Russell. 2020. Peter of Trabibus. In *Encyclopedia of Medieval Philosophy: Philosophy between 500 and 1500*, 1458–1460. Dordrecht: Springer.

Friedman, Russell. Forthcoming. *The Sentences Commentary of Peter of Trabibus.*

Gagliardi, Antonio. 1997. *Guido Cavalcanti e Dante: una questione d'amore*. Catanzaro: Pullano.

Gagliardi, Antonio. 2004. "Species intelligibilis." In *Guido Cavalcanti laico e le origini della poesia europea*, ed. Rossend Arqués, 147–61. Alessandria: Edizioni dell'Orso.

Gallus, Thomas. 2011. *Explanatio in libros Dionysii*, ed. Declan Anthony Lawell. Turnhout: Brepols.

Gallus, Thomas. 2022. *Commentaries on the Angelic Hierarchy*, ed. Declan Anthony Lawell. Turnhout: Brepols.

Gentili, Sonia. 2002. Due definizioni di 'cuore' nel *Convivio* di Dante: 'secreto dentro', 'parte dell'anima e del corpo' (II, 6, 2). *Lettere Italiane* 54 (1): 3–36.

Gentili, Sonia. 2005. *L'uomo aristotelico alle origini della letteratura italiana*. Rome: Carocci.

Gentili, Sonia. 2020. Poesia e filosofia a Firenze tra Santa Croce e Santa Maria Novella. In *The Dominicans and the Making of Florentine Cultural Identity (13th-14th centuries)*, ed. Johannes Bartuschat, Elisa Brilli, and Delphine Carron, 225–41. Florence: Firenze University Press.

Gragnolati, Manuele. 2008. *Experiencing the Afterlife: Soul and Body in Dante and Medieval Culture*. Notre Dame: Notre Dame Press.

Graver, Margaret. 2007. *Stoicism and Emotion*. Chicago: University of Chicago Press.

Graziosi, Elisabetta. 1997. Dante a Cino: Sul cuore di un giurista. *Letture Classensi* 26: 55–91.

Gubbini, Gaia. 2011. 'Radix amoris': Agostino, Dante e Petrarca (con Bernardo di Ventadorn). *Critica del testo* 2: 465–81.

Gubbini, Gaia. 2017. Lust, Desire, Hope. Dante and the Problem of Love (*Inf. V*; *Purg. XVII-XVIII*; *Purg. XXVI*). *Romanistisches Jahrbuch* 68: 173–96.

Hissette, Roland, 1977. *Enquête sur les 219 articles condamnés à Paris le 7 mars 1277*. Louvain: Peeters.

Houser, Rollen Edward. 1979. Transcendental Unity in Petrus de Trabibus. *Franciscan Studies* 39: 49–104.

Huning, Hildebert Alois. 1964–5. Die Stellung des Petrus de Trabibus zur Philosophie, nach dem zweiten Prolog zum ersten Buch seines Sentenzenkommentars Ms. 154, Biblioteca Communale, Assisi. *Franziskanische Studien* 46: 193–286; *Franziskanische Studien* 47: 1–43.

Huning, Hildebert Alois. 1968. The Plurality of Forms According to Petrus de Trabibus O.F.M. *Franciscan Studies* 28: 137–96.

Kent, Bonnie. 1995. *Virtues of the Will: The Transformation of Ethics in the Late Thirteenth Century*. Washington, DC: Catholic University of America Press.

King, Peter. 2008. Aquinas on the Passions. In *Aquinas's Moral Theory: Essays in Honor of Norman Kretzmann*, ed. Scott MacDonald and Eleonore Stump, 101–32. Ithaca: Cornell University Press.

Komline, Han-luen Kantzer. *Augustine on the Will: A Theological Account*. New York: Oxford University Press.

Knuuttila, Simo. 2004. *Emotions in Ancient and Medieval Philosophy*. New York: Oxford University Press.

Ledda, Giuseppe. 2002. *La guerra della lingua: ineffabilità, retorica e narrativa nella Commedia di Dante*. Ravenna: Longo.

Lombardo, Nicholas E. 2011. *The Logic of Desire: Aquinas on Emotion*. Washington, DC: Catholic University of America Press.

Malato, Enrico. 2005. Amor cortese e amor cristiano da Andrea Capellano a Dante. In *Studi su Dante: Lecturae dantis, chiose e altre note dantesche*, 571–657. Cittadella: Bertoncello.

Mazzotta, Giuseppe. 1972. Dante's Literary Typology. *MLN* 87 (1): 1–19.

McIntosh, Mark A. 2021. *The Divine Ideas Tradition in Christian Mystical Theology*. Oxford: Oxford University Press.

Miner, Robert. 2009. *Thomas Aquinas on the Passions: A Study of Summa Theologiae, 1a 2ae 22–48*. Cambridge: Cambridge University Press.

Mocan, Mira. 2018. Intelletto d'amore: la mistica affettiva in Dante. In *Theologus Dantes: Tematiche teologiche nelle opere e nei primi commenti*, ed. Luca Lombardo, Diego Parisi, and Anna Pegoretti, 81–102. Venice: Edizioni Ca' Foscari.

Moevs, Christian. 2005. *The Metaphysics of Dante's Comedy*. New York: Oxford University Press.

Nardi, Bruno. 1944. *Nel mondo di Dante*. Rome: Edizioni di Storia e Letteratura.

Pace, Matteo. 2020. *Ut animalium pictura*: Aristotle's Heart in the Poetry of Giacomo da Lentini. *Traditio* 75: 225–51.

Pasnau, Robert. 2001. *Thomas Aquinas on Human Nature: A Philosophical Study of Summa Theologiae, 1a 75–89*. New York: Cambridge University Press.

Pertile, Lino. 2005. *La punta del disio: semantica del desiderio nella Commedia*. Fiesole: Cadmo.

Peter of Trabibus. *Ordinatio*. Florence, Biblioteca Nazionale Centrale, MS Conventi Soppressi B.V.1149.

Peter of Trabibus. *Quolibetales*. Florence, Biblioteca Nazionale Centrale, MS Conventi Soppressi D.6.359.

Piron, Sylvain. 2008. Le poète et le théologien: une rencontre dans le "studium" de Santa Croce. In *Ut Philosophia Poesis. Questions philosophiques dans l'oeuvre de Dante, Pétrarque et Boccace*, ed. Joël Biard and Fosca Mariani Zini, 73–112. Paris: Vrin.

Porro, Pasquale. 2014. Canto XVIII. Amore e libero arbitrio in Dante. In *Cento canti per cento anni (Lectura Dantis romana): Purgatorio, canti XVIII-XXXIII*, ed. Enrico Malato and Andrea Mazzuchi, 523–60. Rome: Salerno Editrice.

Proto, Enrico. 1915. Note al *Convivio* dantesco: Le ricchezze e la scienza. *Giornale storico della letteratura italiana* 65: 199–262.

Ralby, Rohini. 2021. *Living the Practice: The Way of Love, vol. 1.* Baltimore: Bancroft Press.

Rea, Roberto. 2018. Amor est passio virtutis ymaginative: Immaginazione, immagine e immaginare nella lirica amorosa duecentesca. *Letteratura e arte* 16: 119–37.

Ruini, Roberto. 2007. Un invito alla concordia: il motivo del 'tirare la corda' dai classici al Quattrocento volgare. In *Quattrocento fiorentino e dintorni: saggi di letteratura italiana*, 83–96. Florence: Phasar.

Sasso, Gennaro. 2011. *Dante, Guido e Francesca.* Rome: Viella.

Simoncioli, Feliciano. 1956. *Il problema della libertà umana in Pietro di Giovanni Olivi e Pietro de Trabibus.* Milan: Vita e Pensiero.

Singleton, Charles. 1977. *An Essay on the Vita Nuova.* Baltimore: Johns Hopkins University Press.

Stump, Eleonore. 2011. The Non-Aristotelian Character of Aquinas's Ethics: Aquinas on the Passions. *Faith and Philosophy: Journal of the Society of Christian Philosophers* 28 (1): 29-43.

Tonelli, Natascia. 2015. *Fisiologia della passione. Poesia d'amore e medicina da Cavalcanti a Boccaccio.* Florence: SISMEL-Edizioni del Galluzzo.

Vaura, Tuomas. 2022. Peter de Trabibus on Creation and the Trinity. *Recherches de Théologie et Philosophie Médiévales* 89 (1): 145–95.

Wippel, John. 1977. The Condemnations of 1270 and 1277 at Paris. *The Journal of Medieval and Renaissance Studies* 7: 169–201.

Wippel, John. 1995. Thomas Aquinas and the Condemnation of 1277. *The Modern Schoolman* 72: 233–72.

Wolfson, Harry Austryn. 1935. The Internal Senses in Latin, Arabic, and Hebrew Philosophic Texts. *The Harvard Theological Review* 28 (2): 69–133.

Zambon, Francesco. 2022. Dante e l'intellectus amoris. In *Santi, giullari, romanzieri, poeti: studi per Franco Suitner*, ed. Giuseppe Crimi, Luca Marcozzi, and Anna Pegoretti, 31–42. Ravenna: Longo Editore.

CHAPTER 4

Anatomy of Florence: Illness and Body Politics in the *Commedia*

Andrea Celli

This chapter focuses on several passages in the *Commedia* in which Dante compares Florence to a human body, the injury or illness of which signifies issues affecting the city as a polity. Common in scholastic literature of the thirteenth and fourteenth centuries as well as in Dante, medical allegories of the body politic derive from works of Plato, Aristotle, and Galen, echoed by the vast Greek-Arabic and Latin scholarship of their interpreters.[1] While the *Commedia*'s references to Florence as a human body have generally been interpreted as expressing moral condemnation of greed and arrogance, I would like to suggest a philosophical interpretation. The organic and medical analogies that Dante employs in the poem in relation to Florence hint at a diverse range of failures of the modern city but do not clearly identify the ultimate "source of the ills in the city"

A. Celli (✉)
Department of Literatures, Cultures, and Languages, University of Connecticut, Storrs, CT, USA
e-mail: andrea.celli@uconn.edu

M. Pace (ed.), *Dante and the Sciences of the Human*, Palgrave Studies in Medieval and Early Modern Medicine,
https://doi.org/10.1007/978-3-031-69253-6_4

(*Par.* 16.68). I aim to show that these faults have a single philosophical core.

Despite the central role played by self-governing urban formations such as communes, signorias, and maritime republics in medieval Italy, Dante's reflections on the city appear to be mostly of a moral kind rather than a philosophical one, contrary to the thoroughness of his theological-political view on the empire's soteriological functions. Dante does not belong in the chorus of *ars dictaminis* masters, juridical glossators, and philosophers interested in fostering the role of cities as centers of power and municipal liberties. He never refers to the city as a legitimate body politic but rather mentions—and often condemns—specific cities and their inhabitants. Dante seems to imply that the virtuous functions of the *polis* as discussed by Greek philosophy fully belong to the *monarchia*—the empire—rather than the *città*. In the *Commedia*, Florence, more than any other Italian city, is represented as a nest of greed and is mostly the object of invective.

In only one significant instance does the poem offer an image of Florence as a virtuous city: in *Paradiso* 15, in the Heaven of Mars, when the character Dante encounters his legendary ancestor, Cacciaguida. Many modern commentators have emphasized that Cacciaguida's Florence is mythical and reflects literary *topoi*. Rather than a historically truthful depiction of the city, it serves as a backdrop against which the degeneration of Florence in Dante's time stands out. It is precisely in relation to this contemporary Florence—a reverse image of Cacciaguida's "pure," "sober and chaste" one—that we find in the following canto (*Par.* 16.67–72) one of the few medical analogies for the city.

> The mixing together of persons has always been
> the origin of harm to the city,
> as to you the food that is added on,
> and a blind bull falls sooner
> than a blind lamb, and many times cuts
> more and better one sword than five. (*Par.* 16.67–72)[2]

Critics have noted the gnomic character of the opening lines and the three similes that follow. The cohesiveness of the first tercet is worthy of close examination: "Sempre la confusion de le persone / principio fu del mal de la cittade, / come del vostro il cibo che s'appone" (*Par.* 16.67–69). The opening adverb "sempre" (always, ever) marks the theoretical nature of the statement. Not just about Florence, the principle applies to the city in general ("la cittade"). The "principio" (beginning, origin, but also root cause) of the "mal de la cittade" (the ill or harm to the city) is the "confusion de le persone"—an expression that has been translated in English as the "mingling of the populations" (Mandelbaum 1980) or the "mixing together of persons" (Durling and Martinez 1996), and that could also refer to the "bustle of people" characteristic of a busy city. However, more consistent with the philosophical undertone of this line, we may interpret the term in relation to political treatises such as Giles of Rome's *De Regimine Principum* (between 1277 and 1280), in which the term "confusion" is associated with the absence of a higher authority bringing order to the "multitude."[3] As Brilli (2012a) argues, the denotation of the term "confusion" in Dante is philosophical.

In the same tercet, the following simile further expounds the medical allegory of the body politic: "as to you [living humans] the food that adds on [in the stomach]." It is to be noted that current editions of the *Commedia* (Petrocchi 1994; Sanguineti 2001) make the underlying metaphor of the body politic less evident than other readings of line 67 preferred by many early readers from Iacomo della Lana to Cristoforo Landino. For instance, in two of the three manuscripts of the *Commedia* copied by Boccaccio, we find "come del corpo il cibo che s'appone": "as to the body the food that is added on."[4] "Corpo" in place of "vostro" makes the meaning of the simile immediate and transparent.[5] The metaphor of the body politic becomes explicit. By contrast, numerous implicit terms need to be spelled out to clarify the meaning of "vostro": "as [the origin] of your [harm] [is] the food that is added on [in the stomach]." However, Petrocchi's *lectio difficilior* enjoys broad support today, and it is not my intention to argue against it.

Early readers as well as modern scholars agree on the broadly Aristotelian and scholastic character of the principle uttered here by Cacciaguida.[6] Yet the three similes presented by Cacciaguida in these two tercets remain ambiguous; they hint at slightly different "causes of ill to the city." On the one hand, the source of degeneration is identified with the mixing of different classes of people; on the other, there is the issue of the city's defense, which seems to be better provided by a small and humble, yet distinctly bounded polity ("una spada:" one sword, line 72). Cacciaguida may be implying here that a town a fifth the size of fourteenth-century Florence would be better off when guarded by its virtuous citizens than would a bigger, bustling one (*Par.* 16.48). Is the concern overpopulation and urban drift? Or is it the mixing of people from different social backgrounds and lineages?[7] It would be tempting—but misleading—to interpret the word "confusion" (mixing together) (*Par.* 16.67), along with others such as the antithetical terms "mista" (mixed) and "pura" (pure), used just previously,[8] through the twentieth-century lens of readers familiar with racist and exclusionary political language. Does this passage refer to miscegenation?[9] Or does it point to the need for a hierarchical subordination of the countryside to the city, in accordance with Platonic philosophical conceptions of the *politeia*?[10]

To these different possible nuances, line 69 adds a reference to food piling up in the stomach. The simile evokes images of gluttony and greed, but it can also be read as a reference to the city's disorderly consumerism, typical of mercantile societies, with their imbalanced accumulation of goods. Nicholas Havely (1986) has discussed the significance of the semantic field of hunger as a political metaphor in this and other references to Florence in the *Commedia*. As he shows, gluttony as a sin is a governing theme. In philosophical terms, it is the *pleonexia*, the extreme greed for wealth and material possessions (Balot 2001). Infernal *contrapassi* often entail dreadful images of ingestion. Corporeal figures of speech associated with food and eating tend to be metaphors of rapacity, territorial expansion, envy and self-destructive appetite for power, or partisan divisions. For instance, in *Purg.* 20.73–75, the soul of the French king Hugh Capet conjures up another degrading personification of Florence as a body politic. By synecdoche, the city becomes a belly ("pancia"), which Charles de Valois blows up ("a Fiorenza fa scoppiar la pancia," line 75).[11] "And this comparison is very suitable," says Benvenuto da Imola, "because at that time Florence was very large ('corpulenta'), packed with citizens, swollen with pride."[12] Benvenuto also signals a connection between these tercets and *Inferno* 28—the canto of schismatics—in which the belly of Maometto is split open by a sword, an additional representation of division, or schism, inflicted on the unity of the Christian body politic.[13] Along with other relevant passages in the poem, the association with schism helps to expand the understanding of these corporeal images from the realm of moral judgment to that of political theory.

A third instance in which Florence is compared to a sick body sheds light on what Dante views as the real "origin of ills to the city." In canto 6 of the *Purgatorio*, Dante pronounces a famous apostrophe that begins with the words "Ahi serva Italia." His invective is sparked by witnessing the warm embrace between Virgil and the soul of a troubadour, Sordello.

Hearing from Virgilio the "sweet" name of the city ("terra") of Mantua, in which both poets were born, Sordello feels sudden joy. For a few tercets, Dante digresses from the bliss of the Mantuans' meeting to consider all of Italy, represented as a wretched woman. From the idyllic particularism of Mantua, Dante moves to the sorrow of a destitute homeland. In the last part of the digression, Dante turns back to Italian cities ruled by peasant tyrants before directing his harsh judgment once again toward Florence.

In the last few lines of the canto, Florence is compared to a sick woman, "who finds no rest on her feather bed, / but shields her pain by tossing and turning" (*Purg*. 6.150–51).[14] In the preceding tercets, the poet has—with sarcasm—indicated the nature of the restlessness experienced by the city:

> Athens and Lacedaemon, which made
> the ancient laws and were so civilized,
> made but a little gesture toward right living
> next to you, who make so many subtle
> provisions, that what you spin in October does not
> reach to mid-November.
> How many times, in the period that you remember,
> have you changed laws, coins, offices, and customs,
> and renewed your members! (*Purg*. 6.139–47)[15]

The cause of the illness of contemporary Florence is finally revealed by the comparison to ancient Greek cities. Continuity and stability of the laws is paramount to the "viver bene," the collective welfare, and the well-being of the body politic.[16] The word "membre" (line 147: members, but also limbs), rhyming with "rimembre" (remember), prepares the reader for the simile of the city as the body of a sick woman. As older commentaries on the *Commedia* recognize, the language of this passage is cut from the cloth of political theory.

The use of organic and medical analogies is ubiquitous in medieval political literature in Latin as well as in Greek-Arabic and Jewish scholars of Plato and Aristotle. As an example, it is illuminating to consider the work of Abū Naṣr al-Fārābī, the tenth-century CE forefather of political theory in the Islamic world. Even though Latin translations of some of his works were available to Dante's contemporaries, this author is not thought to be part of Dante's library.[17] However, some scholars have tested readings of Dante's political thought on the backdrop of al-Fārābī's neo-Platonism. In al-Fārābī's treatise on the virtuous city (*Mabādi' Ārā' Ahl al-Madīnah al-Fāḍila*), the reader would easily find examples of medical metaphors of the body politic. Without claiming al-Fārābī as a direct source of Dante, a comparison between the two brings to light a Platonic aspect of Dante's political imagination.[18] For al-Fārābī, however, unlike Dante, the city as a self-governing entity (*al-madīnah*) can function as a virtuous body politic. This conception reflects al-Fārābī's indebtedness to Plato's idea of *kallipolis*, the perfect body politic.

> The most excellent good [*al-khayr al-afḍal*] and the utmost perfection [*al-kamāl al-āqṣā*] is, in the first instance, attained in a city [*madīnah*] ... The excellent city [*al-madīnah al-fāḍila*] resembles the perfect and healthy body, all of whose limbs co-operate to make the life of the animal [*al-ḥayawān*] perfect and to preserve it in this state. (*al-Madīnah al-Fāḍila*, 4.15.3–4).[19]

For Dante, on the contrary, the city cannot be an excellent body politic. Rome is an exception because its name stands for much more than the city itself (Davis 1957). Rome, and Jerusalem, are symbols for the salvific unity of humankind and its well-being, on earth as in heaven. Florence is the reverse image of Rome. One stands for particularism, the other for universalism. Florence represents the city as a separate self-governing body rather than a symbol of unity. The words of Brunetto Latini, whom Dante encounters in *Inferno* 15, further clarify this distinction. By quoting a legend on the origins of the city that exiled them both, he ascribes the internecine strife of contemporary Florence to its original sin: the mixing of Fiesolan and Roman lineages.[20] This belief seems to foreshadow the notion that Cacciaguida will utter in *Par.* 16.67–68: "The mixing together of persons has always been / the origin of harm to the city."

However, on closer examination, Fiesole and Rome should rather be understood here as allegorical designations, referring to two opposite principles. In fact, even after centuries of mixing with the people who literally descended ("discese," *Inf*. 15.62) from mountainous Fiesole, the Roman "sementa santa," the seed of divine election (line 76), can still be discerned within Florence in Dante's time. In the development of this poetic imagery over the space of six *terzine*, Brunetto does not allude to the two distinct principles as bloodlines, as some modern translations and commentaries insinuate, but rather uses disparate images to describe the two lineages: while the people from Fiesole are ultimately described as beasts, Dante's Roman descent is portrayed as a holy plant.[21] This opaque theological-political imagery acquires additional relevance in light of the identity of the speaker, Brunetto Latini, who was an advocate of the ancient liberties of his city.

Is this why Brunetto is condemned in Dante's *Inferno*?[22] Without pretending to answer this question, it must be noted that Brunetto Latini, in the *Tesoretto*, offers a picture of Florence's ambitions and an explanation for the source of its present ills that differ markedly from Dante's own. The intertextual conversation of Dante in the *Commedia* with Brunetto's work is almost explicit. "Be joyous, Florence, you are great indeed, / for over sea and land you beat your wings; / through every part of Hell your name extends!" One may read this damning opening of *Inferno* 26 in light of the following lines from Brunetto's *Tesoretto* to appreciate the difference in attitude toward the standing of the city in the world.

> So that I know none
> Whom I would wish to see
> Have my city
> Entirely in their control,
> or that it be divided;
> But all in common
> Should pull together on a rope
> Of peace and of welfare
> Because a land torn apart
> Cannot survive.
> Truly my heart broke
> With so much sorrow,
> Thinking on the great honor
> And the rich power
> That Florence is used to having
> Almost through the whole world;
> And I, in such anguish,
> Thinking with head downcast,
> Lost the highway,
> And took the crossroad
> Through a strange wood. (*Il Tesoretto*, lines 170–90).[23]

Ultimately, for Brunetto the source of ill in a republic is the elites' lack of education and wisdom.[24] Unlike Dante, he expresses pride in Florence's "grande honore" (great honor) and "la riccha potenza" (rich power). According to Dante, what prevents modern cities from being virtuous self-governing communities is not corruption, the "confusion" brought about by newcomers who recently moved into the city from rural areas, or the greed of uneducated peasants. These are but manifestations of the same fundamental weakness: the inability of the city to reach perfection is linked to its continuous transformation, and its changing nature. Cities are at the mercy of change, as Cacciaguida repeatedly highlights in cantos 15 and 16 of the *Paradiso*.[25] And a correlate of change is that "le cittadi termine hanno:" cities have their end. They do not last forever.

The very success of Florence as a proto-capitalistic society—the Florentine merchant that "cambia e merca" (changes money and trades; *Par.* 16.61)—is bound to the endless transformation to which the city and its members surrender, instead of submitting themselves to the stable rule of a superior, almost everlasting political body, the monarchy and its ancient laws (Caferro 2020). The city is at the mercy of mutation and *fortuna*. The disdain expressed by Cacciaguida toward peasant families who recently moved into Florence to make money should be interpreted in light of this more fundamental issue.[26]

The arrogant Florence of Dante's time is a place of transformation and instability, in contrast to the ideal form of government, the grantor of stability and of that measure of happiness attainable on Earth, the existence of which Dante posits in the treatise on *Monarchia* (Corbett 2020: 67–85). References to Florence as a human body in the *Commedia* should therefore be read in relation to political theory. Maybe Dante is not merely condemning the moral corruption and decadence of modern urban life—in Florence in particular. Rather, the poet is arguing here that the city as a body politic is intrinsically unfit for the attainment of that imperfect earthly form of happiness that he calls *beatitudo huius vitae*.

Notes

1. Shogimen (2012). As relates to Dante, see Ureni (2018). Scholarship often links Dante's membership in the guild of physicians and apothecaries to his philosophical training, on which see Chapter 1 in the present volume.
2. *Par.* 16.67–72: "Sempre la confusion de le persone / principio fu del mal de la cittade, / come del vostro il cibo che s'appone; / e cieco toro più avaccio cade / che cieco agnello; e molte volte taglia / più e meglio una che le cinque spade." In this chapter, I will

quote the Petrocchi edition (1994), and Durling and Martinez's translation of the *Commedia* (1996), with some corrections aimed at increased literalism.

3. This is Elisa Brilli's core argument, in her study of *Paradiso* 16 (2012a: 13–15). See also Brilli (2012b). Vernacular renditions of Giles of Rome's treaty were circulating in late Duecento Florence. See *Il Livro del governamento dei re e dei principi*, Florence, BNCF II.IV.129, volgarizzamento of Giles of Rome's *De Regimine Principum*, 3.2, 16: "Ed in ogne cosa dov'à moltitudine si à confusione, se non v'à alcuno maggiore da cui la moltitudine sia ordinata." See Egidio (2016).
4. Florence, Biblioteca Riccardiana, MS 1035 (Ri), and Toledo, Archivo y Biblioteca Capitulares, MS Zelada 104 6 (To). In Vatican City, Biblioteca Apostolica Vaticana, MS Chigi L.VI 213, that is the latest of three autographs of the *Commedia*, Boccaccio opts for "vostro." See Breschi (2013). Another variant is "ventre," instead of "vostro." See the Sanguineti edition (2001: 465).
5. This variant was preferred by Francesco Torraca (1853–1938) as well. Siro A. Chimenz comments in a footnote in his edition of the poem (3. 770): "La lez. del vostro [male], preferita dal Vandelli e, generalmente, dai commentatori moderni, non sembra attendibile, solo perché toglie evidenza alla contrapposizione con cittade, che è come il corpo di una società, ma soprattutto perché, per se stessa, dà senso meno soddisfacente: vostro male, infatti, non può significare se non 'tutte le vostre malattie', e non tutte le malattie degli uomini derivano da cibo sovrapposto."
6. For instance, Pietro Alighieri mentions Thomas Aquinas, *De Regimine Principum*, 2.3.7. Singleton cites Aristotle, *Politics* 3, 1277b–1278b and 5, 1303a; and Umberto Bosco and Giovanni Reggio refer to Giles of Rome, *De Regimine Principum*, 2.2.11. See results of the query on *Dartmouth Dante Project*, https://dante.dartmouth.edu/search.php. Accessed on 16 March 2023.
7. Brilli (2012a) acutely scrutinizes all these interpretations of the term "confusion" in Dante.
8. *Par.* 16.49–51: "Ma la cittadinanza, ch'è or mista / di Campi, di Certaldo e di Fegghine / pura vediesi ne l'ultimo artista."
9. Francesco Torraca, an Italian philologist who was familiar with the concepts of nineteenth-century "scientific racism," seems to allude to this when he quotes, in relation to Dante's line, a clause

from Frederick II's *Constitutions of Melfi* (1231): the emperor banned marriages with foreigners in the Kingdom of Sicily, because "the mixing of different people had corrupted common decency": "Saepe iam contigisse dolemus, quod post diversarum mixturam gentium passa est ex alienis moribus Regni sinceritas corruptelam. Itaque ex quo homines Regni Siciliae se cum alienigenarum filiis miscuerunt, denigrata est puritas hominum, et excrescente dierum malitia, et infirmitate mentium inquinata conversationibus, ac usibus ceterorum, veritas decrevit in populis, ab illorum ferment grex est fidelium maculatum." Cristoforo Landino thinks that the confusion is caused by the "mescolare," mixing of people who are "di diverse nature, et chostumi."

10. As Abū Naṣr al-Fārābī states (1985: 229–30), "the relation of the village to the city is one of service."
11. The tercet has theological-political undertones, given the association of Florence with the body of Christ, through the "lancia / con la qual giostrò Giuda" (lines 73–74).
12. "Et est propriissima comparatio, quia eo tempore Florentia erat valde corpulenta, plena civibus, inflata superbia. Et iste Carolus scidit eam per ventrem, ita quod fecit inde exire intestina vitalia, scilicet, praecipuos cives, de quorum numero fuit iste praeclarus poeta." I am quoting Benvenuto's *Comentum* (1887) from *Dartmouth Dante Project*, https://dante.dartmouth.edu/search.php. Accessed on 16 March 2023.
13. "Et iste Carolus scidit eam [Florentia] per ventrem, it aquod fecit inde exire intestina vitalia, scilicet, praecipuos cives, de quorum numero fuit iste preclarus poeta." The connection of Florence with the body of Maometto is further stressed by the word "sacco" to indicate the prophet's entrails (*Inf*. 28.26), cut open by the sword of a devil. On the word "sacco" as a technical medical term, see Bertini Malgarini (1989). On *Inferno* 28 and its theological-political language, see Moudarres (2019).
14. *Purg*. 6.149–51: "vedrai te somigliante a quella inferma / che non può trovar posa in su le piume, / ma con dar volta suo dolore scherma."
15. *Purg*. 6.139–47: "Atene e Lacedemona, che fenno / l'antiche leggi e furon sì civili, / fecero al viver bene un picciol cenno / verso di te, che fai tanto sottili / provedimenti, ch'a mezzo novembre / non giugne quel che tu d'ottobre fili. / Quante volte, del tempo

che rimembre, / legge, moneta, officio e costume / hai tu mutato e rinovate membre!"

16. As Benvenuto da Imola paraphrases, "idest, civiliter et beate." Nicola Fosca explains it as "buona organizzazione sociale." Dante seems to focus on the stability of laws rather than on good government.
17. For an abridged introduction to the history of Latin translations of al-Fārābī in the century prior to Dante, see Zonta (2011).
18. For the Arabic tradition of Platonism, see Gutas (2012). For a comparative reading of Dante's *Monarchia* and al-Fārābī's *al-Madīnah al-Fāḍilah*, see Massimo Campanini's introduction to al-Farābī (1996), and Shaw (1995).
19. *al-Madīnah al-Fāḍila*, 4.15.3–4: "Fa-l-khayr^{u} l-afḍal^{u} wa-l-kamāl^{u} al-āqṣā innama yunāl^{u} awwal^{an} bi-l-madīnat^{i} [...] Wa-l-madīnat^{u} l-fāḍilat^{u} tušbih^{u} al-badan^{a} al-tāmm^{a} al-ṣaḥīḥ^{a}, alladhī tata 'āwan^{u} a 'ḍā'uh^{u} kulluha 'alā tarmīm^{i} ḥayāt^{i} al-ḥayawān^{i} wa 'alā ḥafẓihi 'alayhi."
20. As Kantorowicz (1959: 106) noted in a review of Davis's *Dante and the Idea of Rome*, "the tenet of a mixture of noble Roman and base Fiesolan blood in the Florentines, which both Dante and Villani reverted to, may be oral Florentine tradition." Kantorowicz "racial" interpretation of Brunetto's speech is partially reflective of the scholar's biography. In her introduction to Brunetto Latini's *Tesoretto*, Julia Bolton Holloway notes: "Latini adds that the admixture of Catiline Fiesolan with Roman Florentine (for the two groups intermarried) created that discord between Ghibelline and Guelph, and Black and White, that tore apart their beloved Florence, which exiled first Latini and later Dante" (Latini 1981: xvi).
21. Initially, Brunetto's comparison is more coherent and univocal: "tra li lazzi sorbi / si disconvien fruttare al dolce fico" ["among the sour sorbs, the sweet fig is not meant to bear its fruit," *Inf*. 15.65–66].
22. See Skinner (1978: 1. 8–48). Julia Bolton Holloway seems to agree with Richard Kay's claim that "Dante condemns Latini for his Guelph idolatry of Florence as a Republic, rather than being a Ghibelline advocate of Empire" (Latini 1981: xiv).
23. *Il Tesoretto*, lines 170–90 (Latini 1981: 10–13): "Ond'io non so nessuno / Cu'i' volesse vedere / La mia cittade avere / Del tutto a la sua guisa, / Né che fosse divisa; / Ma tutti per comune /

Tirassero una fune / Di pace e di ben fare, / Ché già non può scampare / Terra rotta di parte / Certo lo cor mi parte / Di cotanto dolore, / Pensando il grande honore / E la riccha potenza / Che suole aver fiorenza / Quasi nel mondo tutto; / E io in tal corrocto, / Pensando a capo chino, / Perdei il gran cammino, / E tenni a la traversa / D'una selva diversa."

24. As explained in *La rettorica* (Latini 1968: 2): "Quando io considero li dannaggii del nostro comune e raccolgo nell'animo l'antiche aversitadi delle grandissime cittadi, veggio che non picciola parte di danni v'è messa per uomini molto parlanti sanza sapienza." See Holloway (1988).
25. One might think here of what al-Fārābī's calls the "madīnah al-baddālah," which is the second of four types of cities—"mudun al-ǧāhiliyyah" (the ignorant cities)—, that contrast with the excellent one. "Al-madīnah al-baddālah" could translate to "the changing city" or "the city of constant change." The aim of its population is the acquisition of wealth and riches because wealth is regarded as the main purpose in life.
26. Dante must have been aware that the mythical foundation of Rome, according to Virgil's *Aeneid*, is the result of miscegenation. In Virgil's account, King Latinus is warned by his father Faunus in a dream oracle that his only daughter Lavinia is not to marry a Latin: "Foreign sons-in-law will come, who by blood will raise our name to the stars, and the descendants of their lineage shall witness everything that the Sun gazes at –while traveling back to the other Ocean– fall under their sway" ["externi venient generi, qui sanguine nostrum / nomen in astra ferant quorumque a stirpe nepotes / omnia sub pedibus, qua Sol utrumque recurrens / aspicit Oceanum, vertique regique videbunt;" Virgil, *Aeneid* 7.98–101.] Here "sanguine" seems to be a reference to military prowess and honor: the blood shed in battle. The English translation is mine.

References

Alighieri, Dante. 1963. *La Divina Commedia*, ed. Siro A. Chimenz, 3 vols. Turin: UTET.

Alighieri, Dante. 1980. *The Divine Comedy*, trans. Allen Mandelbaum, 3 vols. Berkeley: University of California Press.

Alighieri, Dante. 1994. *La Commedia secondo l'antica vulgata*, ed. Giorgio Petrocchi, 4 vols. Florence: Le Lettere.

Alighieri, Dante. 1996. *The Divine Comedy of Dante Alighieri*, ed. Robert M. Durling and Ronald L. Martinez, 3 vols. New York and Oxford: Oxford University Press.

Alighieri, Dante. 2018–2020. *Divina Commedia*, ed. Nicola Fosca, 3 vols. Rome: Aracne.

Balot, Ryan K. 2001. *Greed and Injustice in Classical Athens*. Princeton: Princeton University Press.

Benevenuti de Rambaldis de Imola. 1887. *Comentum super Dantis Aldigherij Comoediam*, nunc primum integre in lucem editum sumptibus Guilielmi Warren Vernon, ed. Jacobo Philippo Lacaita. Florence: G. Barbèra.

Bertini Malgarini, Patrizia. 1989. Linguaggio medico e anatomico nelle opere di Dante. *Studi danteschi* 61: 29–108.

Breschi, Giancarlo. 2013. Boccaccio editore della *Commedia*. In *Boccaccio, autore e copista*, ed. Teresa De Robertis, Carla Maria Monti, Marco Petoletti, Giuliano Tanturli, and Stefano Zamponi, 247–53. Florence: Mandragora.

Brilli, Elisa. 2012a. "Ubi multitudo, ibi confusio". Notes sur l'étiologie de la crise de Florence dans Paradis XVI. *Chroniques Italiennes* 22 (1): 1–22.

Brilli, Elisa. 2012b. *Firenze e il profeta: Dante fra teologia e politica*. Rome: Carocci.

Caferro, William. 2020. Dante and Dirittura: Reframing the Florentine Economy. *Dante Studies* 138: 203–218.

Corbett, George. 2020. *Dante's Christian Ethics: Purgatory and Its Moral Contexts*. Cambridge, UK: Cambridge University Press.

Dantis Alagherii Comedia, ed. Federico Sanguineti. 2001. Florence: Edizioni del Galluzzo.

Davis, Charles T. 1957. *Dante and the Idea of Rome*. Oxford, UK: Oxford University Press.

Egidio Romano. 2016. *Il Livro del governamento dei re e dei principi secondo il codice BNCF II.IV.129*, ed. Fiammetta Papi. Pisa: ETS.

Fārābī, Abū Naṣr al-. 1985. *Mabādi' Ārā' Ahl Al-Madīna Al-Fāḍila. Al-Farabi on the Perfect State*, ed. and trans. Richard Walzer. Oxford, UK and New York: Clarendon Press and Oxford University Press.

Fārābī, Abū Naṣr al-. 1996. *La Città Virtuosa*, ed. Massimo Campanini. Milan: Biblioteca universale Rizzoli.

Gutas, Dimitri. 2012. Platon: Tradition arabe. In *Dictionnaire des philosophes antiques*, vol. V-b, ed. Richard Goulet, 854–63. Paris: CNRS.

Havely, Nicholas R. 1986. The Self-consuming City: Florence as Body Politic. *Deutsches Dante-Jahrbuch* 61: 99–114.

Holloway, Julia Bolton. 1988. Chancery and Comedy: Brunetto Latini and Dante Alighieri. *Lectura Dantis* 3: 73–94.

Kantorowicz, Ernst H. 1959. Review of *Dante and the Idea of Rome*, by Charles T. Davis. *Speculum* 34 (1): 103–9.
Latini, Brunetto. 1968. *La Rettorica*, ed. Francesco Maggini. Florence: Le Monnier.
Latini, Brunetto. 1981. *Il Tesoretto (The Little Treasure)*, ed. and trans. Julia Bolton Holloway. New York and London: Garland.
Moudarres, Andrea. 2019. *The Enemy in Italian Renaissance Epic: Images of Hostility from Dante to Tasso*. Newark: University of Delaware Press.
Shaw, Prue. 1995. *Dante, Monarchia*. Cambridge: Cambridge University Press.
Shogimen, Takashi. 2012. Medicine and the Body Politic in Marsilius of Padua's *Defensor pacis*. In *A Companion to Marsilius of Padua*, ed. Gerson Moreno-Riaño and Cary J. Nederman, 71–115. Leiden: Brill.
Skinner, Quentin. 1978. *The Foundations of Modern Political Thought*, 2 vols. Cambridge: Cambridge University Press.
Torraca, Francesco. 2009. *Commento alla Divina Commedia*, ed. Valerio Marucci, 3 vols. Rome: Salerno Editrice.
Ureni, Paola. 2018. Medicine and Dante's Political Thought. In *Dante as Political Theorist: Reading Monarchia*, ed. Maria Luisa Ardizzone, 209–22. Newcastle upon Tyne, UK: Cambridge Scholars Publishing.
Zonta, Mauro. 2011. Al-Fārābī, Latin Translations of. In *Encyclopedia of Medieval Philosophy*, ed. Henrik Lagerlund, 352–54. Dordrecht: Springer.

CHAPTER 5

Neither O nor I: The Handiwork of *Inferno* 24 and 25

Arielle Saiber

It is often observed that the cantos depicting the fate of the Thieves are full of body parts.[1] There are, in fact, more called-out here than anywhere else in *Inferno*, even more than in the cantos that host the unrepentant lustful, gluttonous, violent, and other transgressors we might consider more corporally driven. Among all these anatomical bits and pieces, there is a heightened presence of one in particular—if not always named, certainly implied and certainly unsurprising for those accused of theft—the hand.[2] Now, in Hell, the hands of thieves change from being bound to being incinerated, mutilated, and mutated. When not tied behind their backs by snakes, their hands, together with the bodies to which they belong, are at one moment familiar objects, and in the next, strange things: ashes; stumps on a slithering trunk; lizard-like paws; hands of another sinner; and momentarily nothing. Hands are what Vanni Fucci

A. Saiber (✉)
Department of Modern Languages and Literatures, Johns Hopkins University, Baltimore, MD, USA
e-mail: asaiber1@jhu.edu

M. Pace (ed.), *Dante and the Sciences of the Human*, Palgrave Studies in Medieval and Early Modern Medicine,
https://doi.org/10.1007/978-3-031-69253-6_5

famously uses to throw an obscene curse at God; and hands are what write the very verses in *Inferno* 25 that the Pilgrim/Poet claims surpass those penned by earlier masters of metamorphosis.[3] The hand emerges, in fact, in the opening tercets of *Inferno* 24 through the simile of a "villanello" (line 7) who gazes at a white dusting on his crops that resembles, or rather copies ("assempra," line 4), snow. He then slaps his thigh in frustration, only to realize later that he read the situation incorrectly: the powdery substance was actually "written" by the ink of frost's hand, a short-tempered pen ("ma poco dura a la sua penna tempra," line 6).[4] This image, within the metaphor for the Pilgrim's dismay at seeing Virgil's furrowed brow, quickly resolves into relief, as the cold, pale "words" of worry disappear from Virgil's face. We also see, in these cantos, how the Pilgrim must use his hands to pull himself up ("t'aggrappa," line 29) on the rugged rocks at the base of the Sixth Bolgia to reach the overhang to the Seventh, and that the writing hand is what the Poet feels has become weak after describing the horrific and spectacular "novità" (*Inf.* 25.144) he has penned. These cantos are filled with *maniculae*, directing our gaze in this fiery heap of squirming, knotted bodies, human and reptilian. They point to questions of the visible and invisible, the known and unknown, quantity and nothingness in both the world of objects and things and in that of writing. The main characters I will discuss here are not Vanni, Cacus, or the Florentine thieves, but rather "marks" made by Dante's very hand.

Objects and Things

Hands of all sorts flicker between being *objects* and being *things* throughout *Inferno* 24 and 25, recalling the act of thievery and prompting questions about ownership and identity. Before turning to these cantos, and to *Inf.* 24.100 in particular, however, I would like to offer a few general thoughts about *objects* and *things*. Phenomenology has long explored how to characterize an *object* and a *thing* vis-à-vis how we perceive and use them; I will not rehearse that scholarship here. Philosophers have also long attempted to parse the ways in which the more specified, identifiable, and proprietary *object* differs from a vaguely perceived, unrecognizable, or unknown *thing*. Epistemological and ontological concerns populate such reflections, as do legal, biopolitical, economic, sociological, and psychological questions of ownership. While many are approaches to thinking about the difference between *objects* and *things*, I define an *object* as an entity for which there is a consensus as to its nature. It is something you can point to and say, "that is a house," or "this is a leaf," and most people will agree. The *thing*, on the other hand, exists in an amorphous place of possibility that exceeds or precedes an object. An *object* can lose its status as an object when it estranges or withdraws itself from us, opening itself up to new identifications and interpretations; it then moves into the territory of *thing*. What was once a barn door (an object), for instance, now sits broken on the ground as a slab of wood, prompting a viewer to ask, "what is that?" Its original function is gone, and new uses bubble up as we gaze at it: it could become a table! This *thing* now speaks to us in terms of its form, materiality, age, value, and so forth. The barn door *qua* barn door recedes from our view, and what was previously unseen within it, or about it, begins to speak. While an *object* like a barn door exists independently of human perception and is not ontologically exhausted by its relations with humans or other objects (although Bruno Latour and others might disagree with this point), *things* are *things* because they lack fixed identities. Take, for example, the experience of seeing a tall, dark *thing* in the distance as you drive down a desert highway. As you approach, it comes into focus. It is an oil rig. It is now an identifiable *object*. A *thing* is what an *object* is before we recognize it, name it, and consign its status and utility. A *thing* is also what emerges when we become, for one reason or another, estranged from an *object*, when we see it differently or interact with it differently. *Objects* and *things* are continually alternating with respect to the viewing subject—they depend as much on your knowledge base, perceptive abilities, and imagination as they do on the context in which you interact with them.

Recent scholarship in speculative realism, new materialism, and object-oriented ontology offers numerous ways of thinking about *objects* and *things* in relation to how we perceive them. Latour's Actor-Network Theory (2005) and Graham Harman's challenge to Latour's model are two of the best-known approaches to interrogating the actor-actant, viewer-viewed dynamic. Works in feminist theory, ecocriticism, and animal studies that intersect with the philosophical discourses just noted also add to their considerations the dimensions of empathy, gender, race, sexual orientation, class, economics, and social justice.[5] Ian Bogost, who wrote *Alien Phenomenology: Or What It is Like to Be a Thing* (2012), also co-edits a book series called *Object Lessons*. In each book of the series, an *object* (e.g., a shipping container, rust, a shopping mall, a sock, traffic) takes center stage and an author contemplates its nature through bits of culture, past and present, here and there. The goal of each "object lesson" is to get at its objecthood, but also at its thinginess, as the more you consider an *object* in its fullness, the more complex and estranged it can become. And conversely, some-*thing* as vague as 'sky,' when considered in its fullness, can come into view as an *object*.

A highly influential line of thought in object-oriented ontology has been posed by philosopher Timothy Morton, who challenges us to imagine an entity so massive and so extensively distributed across space and time that in its entirety it is barely locatable and barely comprehensible. Morton calls this entity a "hyperobject" in their 2013 book *Hyperobjects: Philosophy and Ecology at the End of the World*. The whole of a hyperobject dwells outside of our sensory and—in many cases—cognitive reach. As such, I would call it a *thing*. Yet once we know what a given *thing* is, no matter how immense—and Morton's goal is to help us see such things, name them, and know them—it takes on the status of *object*. Climate-change-as-object is Morton's case study: a conglomeration of centuries of human actions and planetary reactions. We see its fingerprint in our weather patterns, migration and extinction patterns, resource scarcity, and our predictions of what the world will experience in the future. It is the radiation and toxins in our bodies. It motivates us to take or not take certain actions. To alter the course of this particularly dangerous hyperobject we must be able to face both the jumble of *things* that produced the hyperobject and the *things* that the hyperobject produces, *things* that, as we come to understand them, become real *objects*—like carbon emissions and melting icecaps—with which we must contend.

Dante did not directly discuss the difference between *objects* and *things* in his works, nor did he elaborate a theory of ownership, not even in the cantos of the Thieves. Yet in *Inferno* 24 and 25, we can see how he reflects on the status of both, that is, on the fluctuation between *objects* and *things* when ownership and identity are in question. He does this by depicting the changes and exchanges with which theft and the three modes of punishing it in Hell present us. Handless snakes and multi-handed reptilian and amphibious creatures feature prominently in the thieves' contrapasso, catalyzing the sinners' various metamorphoses: incineration and resurrection; human-snake fusion; and exchange of form/matter between sinners.[6] Fire, smoke, ashes, and even nothingness accompany these transformations, oscillations between familiar *objects* and unfamiliar *things*.

The Thieving Hand

For Dante's contemporary readers, the Biblical connection between theft, snakes, and original sin was obvious: Adam and Eve "stole" a piece of low-hanging fruit at the encouragement of a cunning snake intent on "stealing" their innocence and bliss. In the Italian Middle Ages, as in Antiquity, theft was thought to be akin to dismemberment, a stolen item as closely linked to its owner as a part of their body. In the writings of Aristotle and in canon law, property was seen as an extension of a person or community; removal of or damage to it, a kind of wound. Dante reflects this understanding in *Inf.* 11.34–36 in relation to violent plundering of goods (what we might think of as armed robbery or mugging), which is punished among the Violent Against Others in the First Round of Circle 7. Suicide, similar to acts of violence against others, lays fierce hands onto a body, albeit one's own, as does profligate spending of one's body of resources; both are punished in Circle 7's Second Round. Blasphemy, sodomy, and usury—yet other forms of violence to a body (God's, nature's, and art's)—are punished in the Third Round. It is worth observing that in the Seventh Circle, the Poet has Virgil clarify to one of the Round's guardians, the centaur Chiron, that neither Virgil nor Dante is a thief (*Inf.* 12.90). Is it a coincidence that thievery is punished in both Circle 7 and Bolgia 7 and that the Commandment "Thou shalt not steal" is the seventh in Augustine's ordering?

Commentators have often discussed why Vanni, who was known for terrible acts of violence, was not placed with the violent-against-others (and their property) in the Seventh Circle. The same question has been asked about the ferocious giant Cacus, who also appears in the Seventh Bolgia, decorated with a veritable den of snakes on his back, and who Dante (inspired by Virgil's word *semihominis* in *Aeneid* 8.193) makes into a centaur, recalling the Seventh Circle's guardians among the violent.[7] Why then, are they here, among the thieves of *Inferno* 24–25? We know from Aquinas, following Aristotle and Augustine, that there are multiple kinds of theft (*ST* 2.2., q.66). There is simple theft, or *overt* robbery, which may be violent but does not, usually, cause irreparable harm (Dante refers to this kind in *Inferno* 11 and again with the Violent Against Others in *Inferno* 12). And there is the *covert* kind of theft, which Aquinas views as causing some sort of deformity to a "body." Covert theft might include the stealing of sacred things, of communal things, or a person—that is, kidnapping. Vanni, Cacus, and the other five Florentines we find in *Inferno* 25 were all guilty of covert theft.

Whether Dante meant to associate each sinner mentioned in these cantos with a single kind of theft has been widely debated by commentators. Most recent scholars have rejected this idea, as well as the subsequent notion that each type of theft would correspond to one of the three types of metamorphosis-punishments that Dante depicts. Also up for debate is whether the thefts that Vanni, Cacus, and the five Florentines committed were only of communal bodies or also of individual ones. Most scholars do agree that the two metamorphoses (that of a human morphing into a snake-lizard and back again, and that of two humans morphing into one another by way of reptiles and back again) Dante claims to be more incredible than anything Lucan or Ovid ever penned *are* the two kinds described in *Inferno* 25.[8] There is no consensus, however, as to whether each thief cycles through all three kinds of transformation (which we might term "trans-material-ization," given the reference to matter in *Inf.* 25.102, or *metousiosis*, as Teodolinda Barolini proposes) or whether each thief experiences only one. Also disputed is why Puccio Sciancato, as well as Cacus, does not morph, or at least does not do so in front of Dante's eyes. In all other cases, however, the contrapasso consists of the sinner being literally robbed of his body and identity by reptiles, fire, and other human bodies.

And so with Vanni Fucci. By stealing sacred objects from Pistoia's Cappella San Jacopo, Vanni stole items from, and mutilated, the body of the church. Now, treacherous hands tied, his own human form (and matter) becomes the site of theft and disfigurement, his body appropriated and estranged. After a bite from a snake at the nape of his neck (*Inf.* 24.97–99), Vanni bursts into flames and then ashes, only to be reborn in human form once again. In *Inferno* 25, other sinners transmute into snakes, lizards, and one another. In all cases, the thieves' whole bodies are embezzled and changed from a familiar *object* (human form) to estranged *things* (fire, ashes, weird reptilian creatures, conglomerations of each other) and even to no-*thing*.

The many perversions, as well as sexualization, of the body in these cantos have been long discussed. The Phoenix's/Christ's resurrected body, Trinity's three-in-one becoming an obscene two-and-one-and-none, the references to alchemy (the sinners of which are punished with diseased, itchy scabs in *Inferno* 29), and the inversion of the natural stages of a fetus's generation and birth, "corrected" by Statius in another twenty-fifth canto,[9] are just a few of these corruptions of bodies and order. While many commentators have noted the perversion of thieving hands in these cantos and how the nature of ownership is connected to that of identity, what has not been considered is how the writing hand, like the thieving hand, drifts between being an *object* and a *thing*, estranging us and educating us, in what it is to "count" in this world.

In what way, then, is writing an *object* or a *thing*? Written letters and words (whether handwritten, typed, printed, engraved, or painted) are what I would consider *objects*: they are graphic marks that have agreed-upon signification and functions. Letters (in alphabet-type languages) have set morphological conventions, even with the presence of countless calligraphic, geometric, historiated, and typographic styles. But what about a verse of poetry, a sentence of prose, or a literary passage—are they *objects*, too? Are they *objects* in terms of their presence on a page, grammatical correctness, and legibility? Are literal interpretations of a given text more *object-like* and metaphorical or allegorical interpretations more *thing-like*? Are writing and reading ever fully "objective"? When writing is crafted to contain multiple levels of meaning—some of which may evade an individual reader and some which may well be unintended—does it enter the realm of *things*, or is it a multilayered *object* that we cannot apprehend in its entirety, like a *hyperobject*? To consider how Dante uses the figure of writing to interrogate the *object-thing* oscillation that occurs with theft brings us into dialogue with questions like these. As we will see, a hand that seeks to gain from others' property or efforts may not only end up empty but may become an agent of its own erasure.

Neither O… nor I

In *Inferno* 24, Dante describes how rapidly Vanni becomes a pile of ash and then returns to human form:

> Né O sì tosto mai né I si scrisse
> com'el s'accese e arse, e cener tutto
> convenne che cascando divenisse.
>   e poi che fu a terra sì distrutto,

la polver si raccolse per sé stessa
e 'n quel medesmo ritornò di butto. (*Inf.* 24.100–105)

Neither *O* nor *I* has ever been written so fast as he
caught fire and burned and was all consumed,
falling, to ashes;
and when he was on the ground, destroyed, the
dust gathered together by itself and instantly became
the same one again.

Since at least as early as L'Ottimo (comm. 1333)[10] scholars have noted that the letters *o* (or *O*) and *i* (or *I*) are quickly—perhaps more quickly than any other letters—written with a quill. Many commentators quote Anonimo Fiorentino's (comm. 1400?) assertion that these letters "si scrivono con un solo tratto di penna" [are written with only a stroke of a pen]. Vanni's transmutation is as fast as lightning, resembling, no doubt, the most expert acts of theft.

From the earliest manuscript copies of the poem to the most recent critical editions, we see instances of both minuscule and majuscule o/O and i/I, italicized and not, and almost no discussion as to how these letters were likely to have been written by the Poet.[11] Charles S. Singleton (comm. 1970–1975) was one of few scholars to take up this matter, arguing that Dante must have been referring to an undotted *i*, and a minuscule one at that, as a majuscule *I* "certainly could not be made with a single stroke of the pen." Lowercase *i*'s were, in fact, rarely (but not never) dotted or marked in some way in Italian scripts of this period, as we can see in many early copies of the *Commedia*.[12] Singleton does not address the "case" of o/O. Many scribes did, however, attempt to set off the *i* and the *o* with lightly drawn slashes (/), the *virgula suspensiva*,[13] as we can see in City of Vatican, Biblioteca Apostolica Vaticana, MS Urbinate Latino 366, f. 41r, and Berlin, Staatsbibliothek zu Berlin Preußischer Kulturbesitz, MS Hamilton 203, f. 22vb.

I have always been surprised by the lack of attention to the form and formation of these two letters. If Dante wrote the letter *i* in an early fourteenth-century Italian gothic or semigothic *littera textualis*, *bastarda*, or a cursive *cancelleresca* or *mercantesca*, and wrote it without a dot (or small line or mark) above the *stem*, that would certainly be a quick, single stroke (even with serifs at the top and bottom), although a minuscule *l* (ell) without an ascending flourish in one of those hands can be rendered even more rapidly than an *i* with a mark. If Dante intended the *I* to be a majuscule, then, as Singleton noted, it would likely have included more strokes: a tall, vertical stem (sometimes two) plus one or two arms or flourishes (with or without serifs) at its base and/or top, and potentially a half crossbar in the middle of the stem or stems. As for the o/O, whether minuscule or majuscule, especially if written with a broad-edge pen tip, this letter takes more effort to create than a c/C, because the o/O is made with two (or more) strokes rather than one.[14]

While we have no examples of Dante's writing, we have an account of it by Leonardo Bruni who, in his *Vita di Dante* (1436), claims to have seen it. Bruni describes Dante's handwriting as "magra e lunga e molto corretta" (38),[15] which sounds more like a quick, light notarial cursive than the slower-to-produce, heavy gothic. Even if this were so, we can only speculate as to the pen tip Dante would have used when writing the *Commedia*, whether angled or pointed, and his choice of script. If Dante used a gothic or semigothic rather than a notarial cursive, a *v* or a *c* would be quicker to write than an *o* or an *I*, as Giovanni Rosadi observes (1917: 18). And speaking of c's, one cannot help but notice the exceptional number of *c*'s in verses 100–102. Still, the question remains: why would Dante choose to use letters (whether o/O and i/I, or even "c's" or "v's" and "l's" that have been long assessed to be o/O and i/I) to describe speed in a canto about theft? A few early commentators, such as Benvenuto (comm. 1375–1380) and L'Anonimo (comm. 1400?), maintained that this speed-simile is original to Dante.[16] Rosadi—the only commentator I have found to have dedicated much thought to *Inf.* 24.100—postulated that Dante may have constructed it based on the medieval Tuscan sayings "non sai fare un' *o*" and "c'è mancato un *i*," as those expressions evoke the letters' simplicity of form (1917: 18). Some scholars have remarked that o/O and i/I recall the simple geometric shapes of the circle and the line, perhaps anticipating Vanni's lewd gesture, thumb prominently displayed between the first and second fingers of a rounded fist (*Inf.* 25.1–3), or the moment when Dante hushes Virgil by placing a finger over his mouth (*Inf.* 25.45), or even, as P. Gioachino Berthier (comm. 1892–97) proposes, the sinner is reduced to a mound of ashes (the o/O) and then, once resurrected, becomes an erect letter *I*. Harold Skulsky (1981: 116) wondered if the two letters were meant to start a cry of pain, *oimé*. Caron Ann Cioffi (1994: 112) saw in them the perversion of one of Zeus's human conquests: the woman Io, turned into a heifer by Hera, but able to communicate her name in the print of her hoof in the sand, resembling the letter *I* surrounded by an *O* (an *iota* and an *omicron* or *omega*). Many readers have commented on the possibility of reading the o/O and the i/I as a reversal of *io,* the first-person subject pronoun, a choice that would reflect how the thieves undergo an eternal undoing of their bodies/identities. Unique to Gabriele Rossetti (comm. 1826–27) are two curious theories: first, that the o/O and the i/I recall the first and last letters of the surname "Orsini," as well as the family's coat of arms, one of its branches decorated with an eel (a creature not dissimilar from the slithering snakes of this bolgia); and second, that the o/O indicates Orsini and i/I Ianicho/Ianiconi, two noble Romans who fought against the Holy Roman Empire.

Whether all, some, or none of these theories matches what Dante had in mind when penning the o/O—i/I simile, it inevitably recalls the hand that writes. We have already mentioned the frost-words that Nature "writes" on a farmer's crops, the frustration "written" on Virgil's face, and Dante's self-congratulatory claim to surpass Ovid and Lucan in inventing metamorphoses (though one could argue that God designed this contrapasso and the Poet is only writing what he saw). But there is more. The first thirty-nine verses of *Inf.* 24 hold a high concentration of homonym (or equivocal) rhymes: tempra:tempra (lines 2, 6); faccia:faccia (lines 11, 13); piglio:piglio (lines 20, 24); and porta:porta (lines 37, 39); and an imperfect one at the conclusion of the canto: dimagra:Di Magra (lines 143, 145). These rhyme words appear, visually and acoustically, to be the same *object* but turn out not to be. They flicker, before our eyes, between one meaning and another—becoming momentarily *things*—just as the "ill-tempered" sinners of this bolgia appear and disappear, burn, turn to ashes, exchange matter and form, all under the Pilgrim's gaze. Dante further invokes writing when he parallels the thieves' transmutations to the way paper turns dark as fire burns it (*Inf.* 25.61–66), and at the end of his time in Seventh Bolgia, he mentions his own pen, which he admits falters amidst all this chaos ("abborra," line 145).

Writing paper, too, is mentioned in these cantos, here leading us to think of ashes and thus of Vanni's (and perhaps the other thieves') contrapasso. After falling into a pile of cinders (echoing the falling frost of the opening verses, or Vanni's "raining down" from Pistoia, *Inf.* 24.122), the Vanni-dust gathers itself together by itself ("la polver si raccolse per sé stessa," *Inf.* 24.104) to reconstitute Vanni's human form. Janet Soskice has called this image one of the saddest in the *Inferno,* as the thief's ashes do not return to Edenic dust and he has no hope of resurrection in heavenly eternity (2017: 34). When Vanni curses God at the opening of *Inferno* 25 and turns once again to ashes, he perversely embodies what Abraham felt after speaking to God (Genesis 18:27), although the patriarch's metaphoric ashes and dust were grounded in humility.

Peter Hawkins described Vanni's contrapasso as an "eternal rewriting of the same hopeless script" (1980: 6). Ashes are, perhaps not coincidently, part and parcel of medieval European writing. While Dante likely wrote his poem using iron-gall ink, he would have known earlier forms of ink—that made from ashes (carbon ink or lamp black)—and he likely used hot ashes to strengthen (temper) his quill or reed tip. In his *lectura* of *Inferno* 24–25 Ambrogio Camozzi Pistoja studies the color white to draw further parallels between ashes and writing: he mentions the frost-snow "writing" simile we saw in the opening of *Inferno* 24; the white paper on which the Poet writes his words (*Inf.* 24.4–6); and the "dust-abacus" (*pulvisculus*) that grounded early Hindu numbers (2021b: 653). He also pairs these "candid" ashes/dust to the dark obfuscations of smoke that cloak the thieves as they transmute.

Writing can be murky. What we think might be a familiar *object* (a word, a letter, a human form) might sometimes shimmer into an unfamiliar *thing*. In the cantos of the Thieves, the written word can be as unstable as the constantly transforming sinners. At times, even the *spoken* word is at risk: silenced by flames, strangulation, or the bites of lizard-like creatures that can change a thief into someone else, the sinners themselves are unable to form words ("parole formar," line 66). Similarly, Vanni's blasphemous sign language—here, the crudest kind of *hand*-writing—is silenced at the beginning of *Inf.* 25. As Anthony Oldcorn (1999) has discussed at length, there are many silencings in these cantos. Covert theft is, of course, enacted with stealth and silence under the cover of darkness, as if protected by "pertugio o elitropia" (*Inf*. 24.93). Keeping in mind that the *visibile parlar* of words on the page (*objects*) can become invisible as words turn into some-*thing* else, we shall now consider a *thing* that I believe has for centuries been hiding in plain sight in these cantos.

Or 0 . . . or 1

In 2010, an undergraduate in my *Commedia* course, Eric Ardolino, asked if the *0* and *1* could be seen not as letters, but instead as numbers. Might these be a *0* and a *1*, mirroring how a person who steals something begins with "zero" of it, and then has "one" of it, leaving the victim subsequently with "zero" of it? And was it a coincidence, he asked, that these two characters happen to be found in verse 100? In hundreds of years of commentary on this canto, no one, to my knowledge, has explored this possibility.[17] Thanks to Ardolino's questions, these two characters now forever flicker in front of me, alternating their identities between something familiar and expected (letters as *objects*) and something unexpected in this verse (numbers as *things*) and back again.

How might Dante have intended us to read the *0* and the *1* in *Inf.* 24.100? If we think of a poet writing in meter, the monosyllabic letters make more immediate sense in a hendecasyllable verse than do the number-words (especially if reading the *mai* as a one-syllable diphthong). To pronounce the *O* as *zero* (or *zefiro, zifero, cifra,* or another thirteenth-century Italian term for this algebraic quantity and place holder) and the *I* as *uno* would add two additional syllables. But zero, especially when used as a placeholder, was also called *nihil* or *nullo* (nothing) in Dante's time, which could be truncated as *nul* or *null'* (Dante uses *null'*, for example, in *Purg.* 1.103 and *Purg.* 7.7–9). The number-word *uno* could be reduced to *un* (he uses *un,* for example, for the number *1* in *Inf.* 25.70). The verse could then be read as "Né *null'* [or *nul*] sì tosto mai né *un* si scrisse," maintaining the eleven beats. While this "null-un" idea might be a stretch, it does offer a neat, palindromic structure *nullun* (or *nulun*), which could be a graphic echo of the thieves' alternations in being. And as long as we are stretching, although perhaps really too far, the Hebrew letter ע (*ayin*) means "nothingness" and also happens to be an anagram of the word for "myself" *aniy* (Seife 2000: 76). Thus, with a flip of letters, the self, a known *object* (*aniy*), becomes the unknown *thing* (*ayin*), and vice versa, just as the *0* and *1* in verse 100 can be inverted to spell *io,* as many readers have proposed. And to stretch further still, the letter *ayin* in Phoenician is drawn as a circle, which is the origin of the Greek letter *o, omicron.* The *omicron* symbol both looks like an empty container and happens to be the first letter of the Greek word οὐδείς, meaning "not at all," "no one," or "nothing." Perhaps Dante knew of the Hebrew *ayin-aniy* anagram through an encounter with the Phoenician *ayin* or Greek *omicron*?

In thinking about the *0*, and related *omicron* and *omega*, one cannot help but to think of *Par.* 26 .17. The *0* of John's Apocalyptic end-time (line 17) in the verse "Alfa e O è di quanta scrittura"[18] is most commonly penned in early copies of *Paradiso* as a majuscule *0* and in some as a minuscule *0*, rather than as Ω or ω. Might it, too, be seen as a zero just as the *alfa* is linked in Hebrew (and in Greek, Arabic, and other languages) to the number one? What is more, in some early manuscripts, the *0* of verse 17 is written as a vertically barred *0*, which resembles a lightly historiated capital *0* but also a diagonally barred zero. Interestingly, in this reference to *Alfa* and *0*, as in *Inf.* 24.100, Dante is also thinking about writing: "Lo ben che fa contenta questa corte / Alfa e O è di quanta scrittura / mi legge Amore o lievemente o forte" [The Good that satisfies this court is Alpha and 0 of all the writing that Love reads to me, whether soft or loud, lines 16–18]. *This* handwriting, however, is that of God.

Let us now look at a small selection of early to mid-fourteenth-century copies of *Inferno* 24.100 to observe how the *0* and *I* were written.[19] (Figs. 5.1–5.7).

Given the hands used—primarily *bastarda cancelleresca* and a *littera textualis*—it is no surprise that the *O* and *I* are in the minuscule. None of these manuscripts, except for Perugia, Biblioteca Comunale Augusta, MS

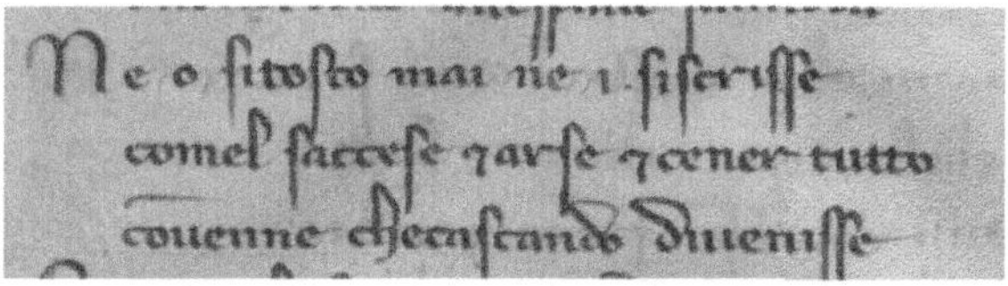

Fig. 5.1 Dante, *Commedia. Inf.* 24.100. Berlin, Staatsbibliothek zu Berlin Preußischer Kulturbesitz, MS Hamilton 203, fol. 20v. Second quarter of the fourteenth century (*Source* https://www.danteonline.it/manoscritti/)

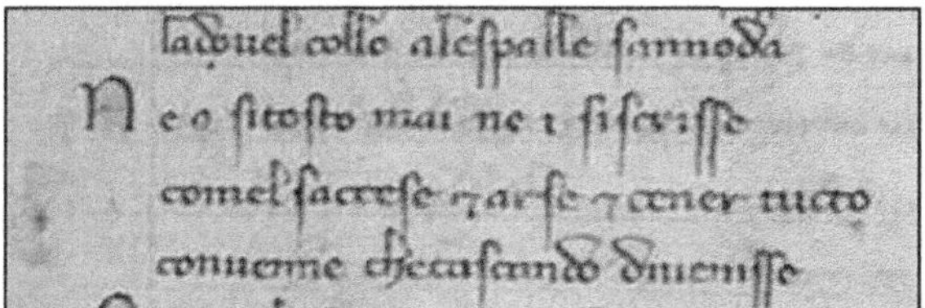

Fig. 5.2 Dante, *Commedia. Inf.* 24.100. Florence, Biblioteca Laurenziana, MS Strozzi 152, fol. 21r. Fourteenth century (*Source* https://www.danteonline.it/manoscritti/)

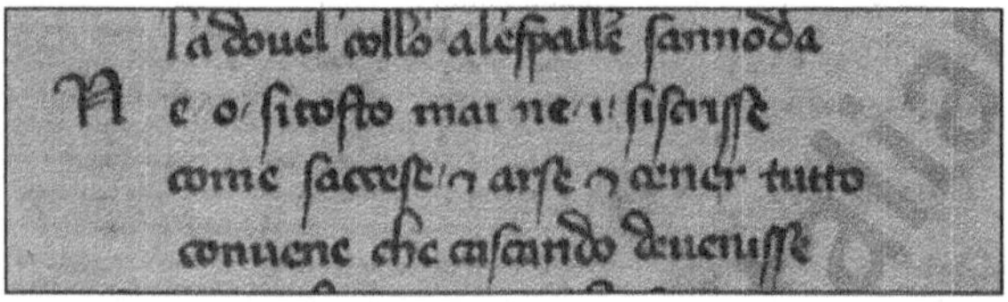

Fig. 5.3 Dante, *Commedia. Inf*. 24.100. Florence, Biblioteca Nazionale Centrale, Fondo Nazionale MS II.I.36, fol. 27r. Fourteenth century (*Source* https://www.danteonline.it/manoscritti/)

Fig. 5.4 Dante, *Commedia. Inf*. 24.100. Florence, Biblioteca Riccardiana, MS 1025, fol. 20v. Fourteenth century (*Source* https://www.danteonline.it/manoscritti/)

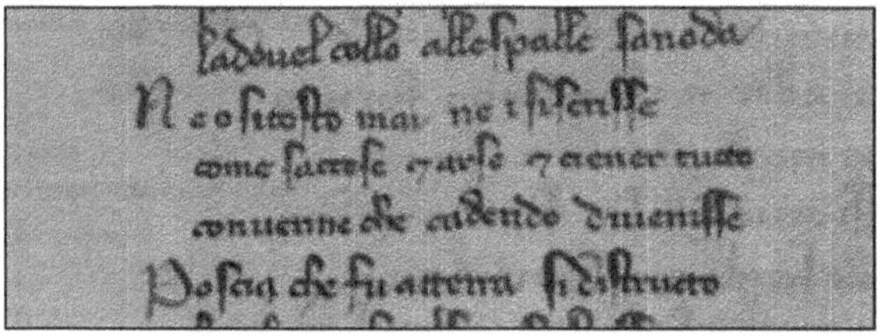

Fig. 5.5 Dante, *Commedia. Inf*. 24.100. Cortona, Biblioteca Comunale e dell'Accademica Etrusca, MS 88, fol. 20v. Fourteenth century, copied by Romolo Lodovici (*Source* https://www.danteonline.it/manoscritti/)

240 (Fig. 5.6), contains dotted *i*'s, and even in MS 240, the *i* in question remains undotted. Does this indicate copyists' inclination to write the *i* more quickly, inspired by Dante's simile? Or might it indicate an almost intuitive pairing of an *i* with a zero, and making the *i* look more like the number one, as zero is becoming more present in Northern Italian culture?

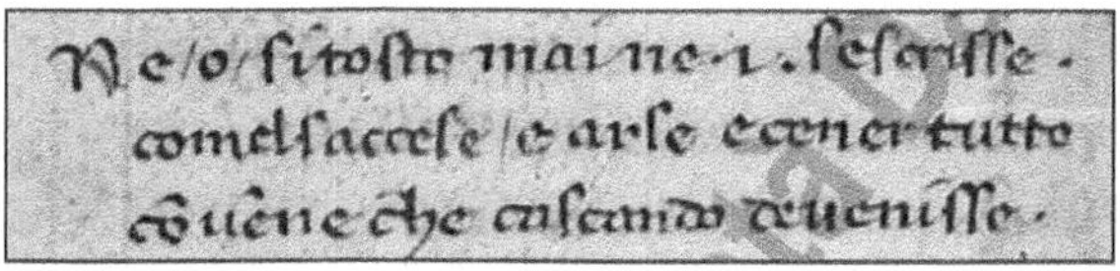

Fig. 5.6 Dante, *Commedia*. *Inf*. 24.100. Perugia, Biblioteca Comunale Augusta, MS 240, fol. 43v. Fourteenth century (*Source* https://www.danteonline.it/manoscritti/)

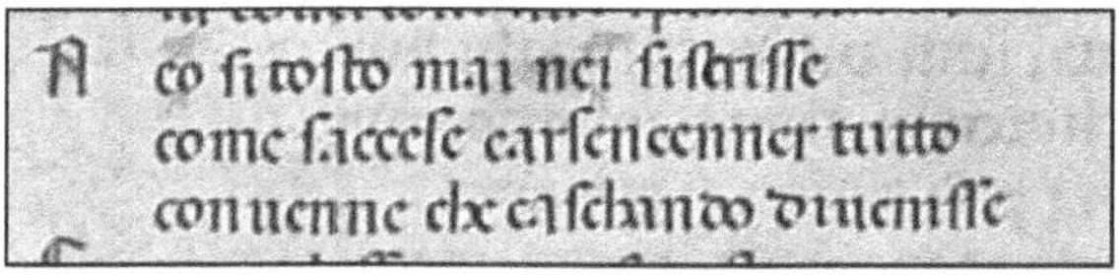

Fig. 5.7 Dante, *Commedia*. *Inf*. 24.100. Perugia, Biblioteca Comunale Augusta, MS L 70 (818), fol. 43r. Fourteenth century (*Source* https://www.danteonline.it/manoscritti/)

If the two objects we are investigating here were, indeed, meant to be read aloud as a *null'* and an *un*, would Dante have written them in a visibly different way? In Dante's Europe, as today, the Hindu-Arabic 0 (zero) and 1 (one) resembled the letters o/O and i/I (especially an undotted, minuscule *i* or majuscule *I*) quite closely, although many instances of zero had a diagonal line through the 0, as is visible in these early thirteenth-century copies of the well-known *algorism* treatises by Johannes de Sacrobosco (Fig. 5.8) and Alexander of Villedieu.

Sacrobosco gives not only the terms *cifra* and *nihil* for zero, but also the Greek *theca* (Θήκη), which means an empty box, hollow cup, or sheath. Karl Menninger (1969: 402–03), in writing about the history of number-words and number-symbols, comments that *theca* (or *teca*) was the name of a medieval branding device used on the face of a convicted criminal.[20] The word *theca* (or *teca*) was also used, and continues to be used, in Christian terminology, to describe a container holding holy relics, the Eucharist, or other sacred objects. Although we have no evidence that the *bestia* of Pistoia was branded with a little circle, Dante places him in the Malebolge (not in Dis for the sin of violent robbery) because he covertly looted sacred objects. His contrapasso empties him out again and again in a loop of annihilation and negation.

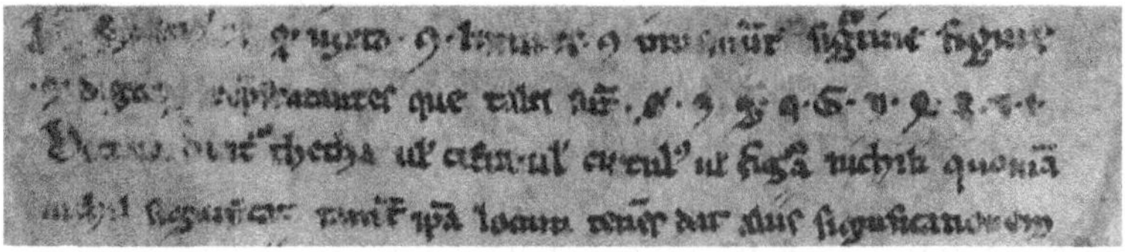

Fig. 5.8 Johannes de Sacrobosco, *Algorismus/De arte numerandi*, c. 1225–1275, Italy. Philadelphia, University of Pennsylvania Library, Lawrence J. Schoenberg Collection, MS 26, fol. 1v: "…sunt nove figure significative nove digitos representantes, que sunt tales 0 9 8 7 6 5 4 3 2 1. Decima figura dicitur thecha [or theca] vel circulus vel cifra vel figura nihili quia nihil significat, sed locum tenes dat aliis significare…" (*Source* https://openn.library.upenn.edu/Data/0001/html/ljs26.html)

Given this association with negation, it is perhaps not surprising, then, that the episode of the Thieves (*Inferno* 24 through the beginning of *Inferno* 26) contains a markedly high incidence of the adverb *non* and other terms of negation (e.g., *nè*, *nessun*, *nulla*, *neente*). In fact, there are more terms of negation among the Thieves than among any other group in the poem. Gabriele Muresu counts forty-two *non*'s between *Inf.* 24.61 and *Inf.* 26.12 (1990: 28–29). Skipping over the most commonly used words in Italian (e.g., definite articles, the conjunction *e*, and the pronoun *che*) *non* appears more frequently than any other word in both *Inferno* 24 and 25.[21] Theft, Dante seems to be saying, wounds communal bodies by negating the value of ownership. As such, the thief's punishment is to become empty of value, empty of *objects*, a zero. The metamorphizing thief will forever be a *cipher*, an ever-changing *thing*.

For all the reasons that we might see a place for zero in the Bolgia of the Thieves, a question still remains: is it plausible that Dante, writing *Inferno* 24 in the early 1300s, would have known of the number zero? As a concept, symbol, and word, *zero* has a long and complicated history. In Dante's time, outside of merchant contexts, zero had not yet been fully accepted as a mathematical object. The number-symbols we use today—1 through 9, that is—first came to be known in Europe through Spain in the late tenth century as versions of Sanskrit letter-numbers, but they would not be widely used in Europe until a few centuries later. Even in the twelfth century, as translations of al-Khwarismi's book on Indian mathematics (which included Hindu-Arabic numerals) began to circulate, the new numbers languished, and zero was still not named or fully integrated into mathematical practice or daily life. Those who did bring the zero into their work developed various notations to signal the concept of शून्य (*sūnya*), the Hindu word for "void" or "nothing": a small circle, a dot, a *theca*, or a blank space. The *zero* begins its life as a known quantity in Europe at the very beginning of the thirteenth century in the widely circulating text by Fibonacci (Leonardo Pisano), *Liber abaci* (1202). Here, he called the "nothing" *cephirum or zephirum* (possibly evoking the light wind of Zephyrus), basing his term on the Arabic (not the Sanskrit) word for "nothing," صفر (*ṣifr*), and he drew the symbol as a small circle, as we can see below (Figs. 5.9, 5.10).

By the early thirteenth century, zero was in use by merchants and some mathematicians, but it was still a suspicious entity, often associated with the demonic: a dangerous idea coming from the Muslim world. Here, I cannot help but think of the analogy Dante makes in *Inf.* 24.112–118 between Vanni's reduction to ashes and a demonic possession or

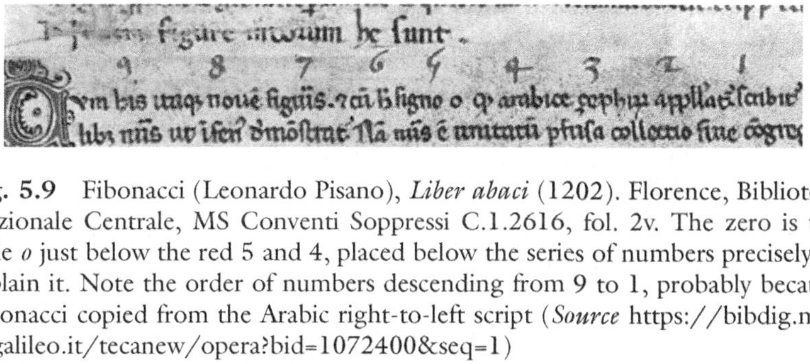

Fig. 5.9 Fibonacci (Leonardo Pisano), *Liber abaci* (1202). Florence, Biblioteca Nazionale Centrale, MS Conventi Soppressi C.1.2616, fol. 2v. The zero is the little *o* just below the red 5 and 4, placed below the series of numbers precisely to explain it. Note the order of numbers descending from 9 to 1, probably because Fibonacci copied from the Arabic right-to-left script (*Source* https://bibdig.museogalileo.it/tecanew/opera?bid=1072400&seq=1)

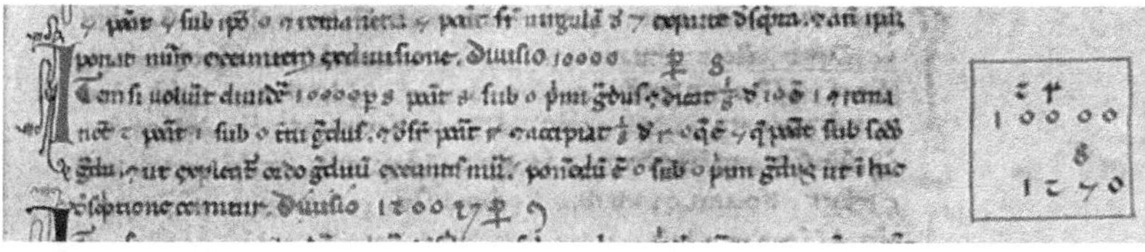

Fig. 5.10 Fibonacci (Leonardo Pisano), Liber abaci (1202). Florence, Biblioteca Nazionale Centrale, MS Conventi Soppressi C.1.2616, fol. 13r (*Source* https://bibdig.museogalileo.it/tecanew/opera?bid=1072400&seq=1)

some sort of "occlusion" ("d'altra oppilazion," line 114)—perhaps like an epileptic seizure, as many scholars have suggested—that pulls a person down to the ground and renders them unconscious.

By Dante's time, 1307 to be precise, the zero is becoming more popular in Europe. Jacopo da Firenze, in Montpellier, wrote the popular *Tractatus Algorismi*, which describes zero as signifying nothing, but having "the power to make signify" (2.2) and calls it *zeuero* and even, in one manuscript copy, *zero* (Høyrup 2007: 50–53).[22] While his claim is still up for debate, historian of mathematics Jens Høyrup identifies Jacopo's *Tractatus* as Europe's earliest vernacular algebra treatise known to date.[23] In the section introducing the Hindu-Arab numerals, on fol. 2r of the fifteenth-century manuscript copy Høyrup translated and edited (City of Vatican, Biblioteca Apostolica Vaticana, MS Vaticano Latino 4826), the *zero* is named and looks, in fact, like a small circle.

Yet even with this progress, zero was still a new *thing* in Dante's time. Resistance to using Hindu-Arabic numerals coupled with the erratic adoption of standardized symbols and names for numbers meant that many mathematical treatises in the thirteenth century still used Roman numerals or wrote numbers (as well as operators) as words. But there was yet another reason why Europeans were slow to adopt the Hindu-Arabic numerals: these new graphic marks for quantities were easily manipulated, leading to miscalculations, at times unintentional and at times quite intentional. Menninger notes that in 1299, the Florentine government issued an ordinance making it illegal to use Hindu-Arabic numerals in accounting books, because it was too easy to scribble one number and pretend it was meant to be another. A 6 or a 9, for example, might be scrawled such that it could look like a 0, or vice versa (1969: 427).[24] Such calligraphic confusion certainly parallels the confusion of the metamorphoses in *Inferno* 24 and 25. The characters o/O and i/I in *Inf.* 24.100, in the several copies we see above, morph before our eyes: are they letters or are they numbers?

These manuscript samples are not, of course, in Dante's hand, and the chronology of adoption of Hindu-Arabic numerals is such that we cannot know for certain how well Dante knew of the number zero. But we can say that by the time these later fourteenth-century copies were being made, the scribes, especially those in contact with the merchant community, would have known of it. Did *they* consider Dante's o/O and the i/I to be a *0* and a *1*? What might *their* hand have erased or added to Dante's? While we cannot say that Dante intended for us to see the o/

O and the i/I as fluctuating between letters and numbers, nor whether scribes saw them that way, such a fluctuation does parallel the thieves' identity switch between an *objective*ly human form and a series of other *things.* It also parallels the concepts of going from having something (1) to having nothing (0), from being something (1) to being nothing (0). This flicker also gives us a glimpse of *no-things,* that is, moments in which the thief is neither one thing nor another (neither a person, nor a reptile, nor flames, nor ashes), nor is he fully hybrid or fully himself ("né due ne uno," *Inf.* 25.69). The o/O and i/I a-*void* being fixed *objects*, yet they are not, I would argue, "empty letters" as James Chiampi proposed (1984: 56). They are handmade *things* bursting with multiple meanings, much like the enigmatic number-words of *Purg.* 33.43: "un cinquecento diece e cinque" [a five hundred ten and five]. What is more, as we will see in the final section of this essay, the o/O and i/I link not only to the status of "having" (or not) and "being" (or not) in these cantos, but to "doing" (or not).

Vestigial Marks

As Dante and Virgil climb their way out of the bridgeless Sixth Bolgia to reach a vantage point overlooking the Seventh, the Pilgrim begins to lose steam (*Inf.* 24.19–63). Virgil prods an exhausted Pilgrim by dangling before him the promise of fame ("fama," line 48). While this *fama* might be the same worldly fame that Brunetto championed in *Inferno* 15, neither Pilgrim nor Poet, of course, should aim for that. Dante should instead be motivated to strive for the *fama* that results from worthy endeavors, such as scaling the rocks in front of him, a mountain later that brings him to the Divine, and writing about it.[25] Such efforts are sure to make lasting impacts, not simply an ephemeral mark on the ground, a "vestigio in terra" (line 50). The lazy person's life, Virgil warns, will result in little more than transient "smoke in air and foam on water" ("fummo in aere e in acqua la schiuma," line 51).

The term *vestigio* denotes a trace of something that remains, but no longer has use or value, something destined to disappear, like a footprint in the sand/snow. Thievery is by nature vestigial. While it might damage a body or bodies, it barely leaves a mark in the sense Virgil intends here. It takes the quick and easy path to doing, to achieving, something. It appropriates what was made (produced, earned) by someone else, and accomplishes nothing of value, and creates nothing of permanence. We know Dante was impatient with those who did not aspire toward achieving knowledge or contributing works of value (see, for example, *Conv.* 11.7–8 and *Conv.* 4.5.9). It is not surprising, then, that a thief, Vanni Fucci, would be one of the few sinners in Dante's hell to delight in broadcasting his "bestial" (line 24) crimes—boasting not only that he stole church appointments, but also that he falsely blamed his action on someone else (and of his many other violent escapades). Clearly, he continues to attempt to leave indelible marks, not just the useless, fleeting ones that ultimately led him to Hell. He feels shame ("vergogna," line 132) that Dante has caught him ("colto," line 133) not in his guilt, but in his marklessness, his emptiness, his zero-ness. The proudest sinner against God who Dante encounters in Hell ("spirto in Dio tanto superbo," *Inf.* 25.15) even tries to leave a mark *on* Dante, with a cruel prophecy about the Poet's and Florence's political future tribulations (*Inf.* 24.142–151) and then, as if that was not bad enough, *on* God Himself, with his double-fisted curse (*Inf.* 25.1–3). Yet his marks are nothing more than vestiges, words wrapped in "hot wind" ("vapor," *Inf.* 24.145) and "turbid clouds" ("torbidi nuvoli," *Inf.* 24.146), the very "smoke" and "foam" about which Virgil warned earlier in the canto. Vanni's words initially make tumultuous marks, but ultimately, they leave nothing more than a light scar, a smudge on the world. They leave him empty-handed.

The empty-handedness of theft may have, in fact, reminded Dante of gambling, an act that is also easily done but makes nothing of value and often leaves a person with empty pockets. While gambling is not mentioned in these cantos, nor has a possible parallel between thievery and gambling been discussed by critics, a brief look at a medieval dice game might reveal a connection. At the opening of *Purg.* 6.1–12, Dante mentions *zara*.[26] One thing we know about this game is that a bad dice roll (above or below a certain number) will prompt the roller's competitors to call out "zara!," indicating that the roller has gained no points and gets nothing. The word *zara*, related to *azzardo* (hazard, chance), seems to have come to Italy from the Egyptian Arabic (or Ottoman Turkish) term *zahr* (زهر for dice).[27] At least one early commentator, Francesco da Buti (comm. 1385–95), associated the word *zara* with the word *zero*. He wrote that the cry "zara!" implies "nulla, come zero nell'Abbaco" [nothing, as the zero in the abacus]. There appears to be no etymological link between *zara* and *zero*, but given the similarities in the words' meaning and their close spelling in Italian, it is not surprising to find the two words correlated in fourteenth-century Italy, as it might have been in Dante's mind.

Theft and gambling rob us of the chance to make a true mark on the world, and as such, lead to nothing. As Giambattista Vico wrote, *verum ipsum factum*, the true is the made. By the end of his time in the gritty, sandy "zavorra" (from the Latin *saburra, Inf.* 25.42) of the Thieves, watching as these perverse metamorphoses leave only vestiges of humanity, the Poet feels confused ("occhi miei confusi," line 145) and dismayed ("animo smagato," line 146). He excuses himself if his pen falters ("la penna abborra," line 144), as the novelty ("novità," line 144) of it all is overwhelming. While God's hand, as the allegory would have it, created this contrapasso, Dante's hand has done the work of taking it all down, and he is clearly exhausted from his scribal efforts.

There is much more to research and to say about the topics we have introduced here. What, for example, are the consequences of theft's harm to bodies (human, religious, civic, and thief's own)? What data can we find to determine Dante's familiarity with the number zero? What would have been the most likely calligraphic form and pen tip used by Dante, and how would he have written letters and Hindu-Arabic numbers, if he wrote them at all? How much would copyists' hands change (metamorphize!) Dante's own?

Né O sì tosto mai né I si scrisse challenges the reader to see differently. Commenting on the verses appearing just before *Inf.* 24.100, Attilio Momigliano (comm. 1946–51) wrote that "qui e nel Canto XXV la parola è un perenne disegno in movimento, un disegno drammatico che si svolge in mezzo a un magico silenzio" [here and in Canto 25 the word is a perennial design in motion, a dramatic design that unfolds within a magical silence]. When our eyes see marks on the page vacillate between letters and numbers, when we think about theft from the perspective of making or not making lasting marks—that is being an enduring *object* or a vestigial *thing*—we see the Poet's expert handiwork. The cantos of the Thieves, especially two of its characters, ask us to contemplate what possession and dispossession entail, what is truly ours and what is not, even as we gaze at something as familiar as our own hand and the ciphers it writes.

Notes

1. I presented an earlier version of this essay as a talk at the Dante Society of America's annual conference *Visibile Parlare and the Invisible* at the University of Toronto on May 4, 2019. I would like to acknowledge Eric Ardolino, a Romance Languages and Chemistry double-major at Bowdoin College who took my *Commedia* course in Spring 2010. It was his observation on *Inf.* 24.100 that catalyzed this current project. I would also like to thank Kenneth P. Clarke for his valuable insights on an earlier version of this essay.
2. The word *mano* and its variations occur most frequently, however, not in the cantos of the thieves, but in *Inf.* 13 (4x), *Inf.* 28 (4x), and in *Purg.* 3 and *Purg.* 27 (4x).
3. Many are the discussions of the classical and biblical source material for Dante's metamorphoses, which I will not discuss here. Studies on gesture, such as that by Vanni Fucci at the opening of *Inferno* 25, are less frequent. For an analysis of hand movements and communications in these cantos and throughout Dante's works, see Webb (2022).
4. All citations in Italian of the *Commedia* are from the Petrocchi edition (Alighieri 1966–67). Translations, except where noted, are from the Durling-Martinez edition (Alighieri 1996–2010).
5. For an ample bibliography on these approaches, see my article (Saiber 2020), especially notes 7 and 8.
6. See especially Teodolinda Barolini (2018) on the question of form versus matter in Dante, on the body, the self, and the "forma vera" in Dante's *Piangete amanti* (*VN* 8) and *Inferno* 24.
7. Whether Cacus is the guardian of the Seventh Bolgia or merely a sinner is unclear.
8. See Livraghi (2017).
9. For a discussion of the thieves' metamorphoses as a parody of coition, semination, gestation, birth, growth, death, and resurrection in *Purg.* 25, see, for example, Shapiro (1990). A number of scholars have also noted that the rhyme *bocca, tocca, scocca* of *Inf.* 25.92–96 recurs in a different order in *Purg.* 25.17–21. See, for example, Oldcorn (1999: 342).
10. All commentaries cited with "comm." and a date in parentheses are from the Dartmouth Dante Project: http://dantelab.dartmouth.edu.

11. Most twentieth-century critical editions of the *Commedia* capitalize the O and the I, including Giorgio Inglese's 2021 *edizione nazionale* for the Società Dantesca Italiana. Giuseppe Campi (1888–1893) is one of the few commentators to note the variation between miniscule and majuscule letters in manuscript copies. No one has discussed editors' decisions regarding capitalizing or italicizing the letters in print editions.
12. See the many scanned early manuscripts online curated by the Società Dantesca: https://www.danteonline.it/manoscritti/ and Prue Shaw's collation of seven early manuscripts of the *Commedia*, online at https://www.dantecommedia.it. Regarding the Italian hands in Dante's time and just after, see, for example, Casamassima (1988), Bertelli (2011–2016), the many samples in Boschi Rotiroti (2004), Ceccherini (2010a, 2010b, 2010c,2021), Ceccherini and De Robertis (2015), De Robertis (2010, 2012, 2021).
13. On the *virgula suspensiva,* see Parkes (1992).
14. Campi's commentary cites an observation by a certain "Romani" that a "c" is quicker to make than an "o," but gives no bibliographic details or elaboration. Francesco Torraca (comm. 1905) believes that the speed of a writing an o/O and an i/I (he does not specify regarding capitalization) in the semigothic of Dante's time would make the simile even more apparent; I would argue the opposite, given the number of strokes it takes to make o/O and i/I; Charles Grandgent (comm. 1909–1913) mentions that the *i* must be an undotted one, but makes no mention of capitalization and does not postulate the calligraphic hand.
15. Thank you to Kenneth P. Clarke for referring me to Berté-Fiorilla-Chiodo-Valente (2017: 239), where Berté and Fiorilla discuss this passage from Bruni in their Introduction. They cite Manlio Pastore Stocchi's reading of the Bruni passage (2012: 136). See also Petrucci (2017: 161).
16. Luigi Pietrobono (comm. 1949) and Manfredi Porena (comm. 1946–1948) also attributed this to Dante.
17. I thank Camozzi Pistoja (2021a, 2021b), who was in attendance when I first discussed this theory in 2019 at the Dante Society of America Conference in Toronto, for having cited my forthcoming research in his *lectura dantis* of *Inf.* 25 (not *Inf.* 24, where the verse appears).

18. Robert Durling and Ronald Martinez, in their commentary on this verse note that the *omega* was rarely written out as a word in medieval Latin bibles. See Alighieri (1996–2010: 3, 527).
19. I consulted many fourteenth-century copies available on the Società Dantesca's website and the seven available on Prue Shaw's digital edition of the *Commedia*. Of these, I chose just a few of clearest writing to include here. In all cases, the o/O and the i/I are in lowercase.
20. Menninger does not provide a source for this association claim.
21. I thank my student Lorenzo Hess (Bowdoin, '23) for creating a computer algorithm that gave us this data.
22. On Jacopo da Firenze and the earliest vernacular algebra, see also Høyrup (2006), and Karpinski (1929).
23. Menso Folkerts (2011) suggests that the section on algebra in the manuscript copy that Høyrup studies may have been added by the scribe and that Jacopo's original 1307 treatise might not have contained this section at all. I have not yet been able to examine the presence (or calligraphic traits) of the zero in the three manuscripts.
24. Menninger does not give a source for this claim. See also Raffa (1992: 128–29), who cites Menninger's finding and considers Dante's knowledge of Hindu-Arabic numbers.
25. See Robin Kirkpatrick (1981) for a discussion of the importance of labor in *Inferno* 24 and 25.
26. See the *TLIO* on *zara*. See also Zdekauer (1886).
27. Thank you to Robert Morrison for help tracing the origin of this word.

Manuscripts

Berlin, Staatsbibliothek zu Berlin Preußischer Kulturbesitz, MS Hamilton 203 (Boschi Rotiroti 9, Roddewig n. 16).

City of Vatican, Biblioteca Apostolica Vaticana, MS Urbinate Latino 366 (Boschi Rotiroti 114, Roddewig n. 691)

Cortona, Biblioteca Comunale e dell'Accademia Etrusca, MS 88 (Boschi Rotiroti 54, Roddewig n. 76).

Florence, Biblioteca Medicea Laurenziana, MS Strozzi 152 (Boschi Rotiroti 108, Roddewig n. 204).

Florence, Biblioteca Nazionale Centrale, MS II.I.36 (Boschi Rotiroti 119, Roddewig n. 231).

Florence, Biblioteca Riccardiana, MS 1025. (Boschi Rotiroti 151, Roddewig n. 312).

Florence, Biblioteca Nazionale Centrale, MS Conv. Soppr. V.1.2616.

Perugia, Biblioteca Comunale Augusta, MS D.58 (240) (Boschi Rotiroti 241, Roddewig n. 601).

Philadelphia, University of Pennsylvania Library, Lawrence J. Schoenberg Collection, MS 26.

Sitography

Dartmouth Dante Project. Dante Lab, http://dantelab.dartmouth.edu. Accessed 20 June 2024.

Shaw, Prue, ed. 2021. *Dante Alighieri* Commedia. *A Digital Edition*. Second Edition. Florence: Fondazione Ezio Franceschini/Saskatoon: Inkless Editions. https://www.dantecommedia.it. Accessed 20 June 2024.

Società Dantesca Italiana. https://www.danteonline.it/manoscritti/. Accessed 20 June 2024.

TLIO. Tesoro della Lingua Italiana delle Origini. http://tlio.ovi.cnr.it/TLIO/. Accessed 20 June 2024.

References

Alighieri, Dante. 1966–1967. *La Commedia secondo l'antica vulgata*, ed. Giorgio Petrocchi, 2nd repr. corr. ed., 4 vols. Florence: Le Lettere, 1994; orig. Milan: Mondadori.

Alighieri, Dante. 1996–2010. *The Divine Comedy*, trans. and ed. by Robert Durling, ed. Ronald Martinez. Oxford: Oxford University Press.

Barolini, Teodolinda. 2018. *Inferno* 24: Metamorphosis (Ovid). *Commento Baroliniano*, Digital Dante. New York, NY: Columbia University Libraries. https://digitaldante.columbia.edu/dante/divine-comedy/inferno/inferno-24/. Accessed 10 August 2023.

Berté, Monica, Maurizio Fiorilla, Sonia Chiodo, and Isabella Valente, eds. 2017. *Dante Alighieri, Opere di dubbia attribuzione e altri documenti danteschi: Le Vite di Dante dal XIV al XVI secolo. Iconografia dantesca.*, vol. 4. Rome: Salerno Editrice.

Bertelli, Sandro. 2011–2016. *La tradizione della "Commedia" dai manoscritti al testo.* 1. *I codici trecenteschi (entro l'antica vulgata) conservati a Firenze*; 2: *I codici trecenteschi (oltre l'antica vulgata) conservati a Firenze.* Florence: L.S. Olschki.

Bogost, Ian. 2012. *Alien Phenomenology, or What It's Like to Be a Thing*. Minneapolis: University of Minnesota Press.

Boschi Rotiroti, Marisa. 2004. *Codicologia trecentesca della* Commedia*: entro e oltre l'antica vulgate*. Rome: Viella.

Camozzi Pistoja, Ambrogio. 2021a. Lettura e interpretazione del canto XXIV. In *Voci sull'*Inferno *di Dante,* eds. Zygmunt G. Barański and Maria Antonietta Terzoli, 623–648. Rome: Carocci.

Camozzi Pistoja, Ambrogio, 2021b. Lettura e interpretazione del canto XXV. In *Voci sull'*Inferno *di Dante,* eds. by Zygmunt G. Barański and Maria Antonietta Terzoli, 649–670. Rome: Carocci.

Casamassima, Emanuele. 1988. *Tradizione corsiva e tradizione libraria nella scrittura latina del Medioevo*. Rome: Vecchiarelli.

Ceccherini, Irene. 2010a. La cultura grafica di Andrea Lancia. *Rivista di Studi danteschi* 10 (2): 351-367

Ceccherini, Irene. 2010b. Le scritture dei notai e dei mercanti a Firenze tra Duecento e Trecento: unità, varietà, stile. *Medioevo e Rinascimento*. 24 (21): 29–68.

Ceccherini, Irene. 2010c. Teaching, Function and Social Diffusion of Writing in Thirteenth and Fourteenth-Century Florence. In *Teaching Writing, Learning to Write: Proceedings of the XVI Colloquium of the Comité international de paléographie latine, held at the Institute of English Studies (University of London, 2–5 September 2008)*, ed. Pamela Robinson, 177–92. London: King's College London, Centre for Late Antique & Medieval Studies.

Ceccherini, Irene. 2021. Uno "scriptorium" diffuso: copisti e notai. In *Onorevole e antico cittadino di Firenze. Il Bargello per Dante*, eds. Luca Azzetta, Sonia Chiodo, and Teresa De Robertis, 203-07. Florence: Mandragora.

Ceccherini, Irene, and Teresa De Robertis. 2015. Scriptoria e cancellerie nella Firenze del XIV secolo. In *Scriptorium. Wesen, Funktion, Eigenheiten. Comité international de paléographie latine, 18. Kolloquium (St. Gallen 11–14 September 2013)*, eds. Andreas Nievergelt et al. 141–69. Munich: Bayerische Akademie der Wissenschaften in Kommission bei Verlag C.H. Beck.

Chiampi, James T. 1984. The Fate of Writing: The Punishment of the Thieves in the *Inferno*. *Dante Studies* 102: 51–60.

Cioffi, Caron Ann. 1994. The Anxieties of Ovidian Influence: Theft in *Inferno* 24 and 25. *Dante Studies* 112: 77–100.

De Robertis, Teresa. 2010. Scritture di libri, scritture di notai. *Medioevo e Rinascimento* 24:1-27.

De Robertis, Teresa. 2012. Digrafia nel Trecento: Andrea Lancia e Francesco di ser Nardo da Barberino. *Medioevo e Rinascimento* 26 (23): 221–35.

De Robertis, Teresa. 2021. Dante come libro. In *Onorevole e antico cittadino di Firenze. Il Bargello per Dante*, ed. Luca Azzetta, Sonia Chiodo, and Teresa De Robertis, 79–87. Florence: Mandragora.

Folkerts, Menso. 2011. Review of Jens Høyrup. *Jacopo da Firenze's* Tractatus Algorismi *and Early Italian Abbacus Culture*. Basel: Birkhäuser, 2007. *Annals of Science* 68: 282–284.

Hawkins, Peter. 1980. Virtuosity and Virtue: Poetic Self-Reflection in the *Commedia*. *Dante Studies* 98: 1–18.

Høyrup, Jens. 2006. Jacopo da Firenze and the Beginning of Italian Vernacular Algebra. *Historia Mathematica* 33 (1): 4–42.

Høyrup, Jens. 2007. *Jacopo da Firenze's* Tractatus Algorismi *and Early Italian Abbacus Culture*. Basel: Birkhäuser.

Karpinski, Louis Charles. 1929. The Italian Arithmetic and Algebra of Master Jacob of Florence, 1307. *Archeion* 11 (2–3): 170–177.

Kirkpatrick, Robin. 1981. Inferno XXV. In *Cambridge Readings in Dante's Comedy*, ed. Kenelm Foster and Peter Boyde, 23–48. Cambridge: Cambridge University Press.

Latour, Bruno. 2005. *Reassembling the Social: An Introduction to Actor-Network-Theory*. New York: Oxford University Press.

Livraghi, Leyla. 2017. Sabello e Nasidio, Cadmo e Aretusa. *Dante: Rivista Internazionale di Studi su Dante Alighieri* 14: 55–66.

Menninger, Karl. 1969. *Number Words and Number Symbols. Cultural History of Numbers* (1958). Trans. Paul Broneer. Cambridge: M.I.T. Press.

Morton, Timothy. 2013. *Hyperobjects: Philosophy and Ecology at the End of the World*. Minnesota: University of Minnesota Press.

Muresu, Gabriele. 1990. I ladri di Malebolge (*Inferno* XXIV-XXV). In *I ladri di Malebolge: Saggi di semantica dantesca*, 9–50. Rome: Bulzoni.

Oldcorn, Anthony. 1999. The Perverse Image. In *Lectura Dantis: Inferno*, ed. Allen Mandelbaum, Anthony Oldcorn, and Charles Ross, 328–347. Berkeley: University of California Press.

Parkes, Malcolm B. 1992. *Pause and Effect: An Introduction to the History of Punctuation in the West*. Aldershot: Scholar Press.

Pastore Stocchi, Manlio, ed. 2012. *Dante Alighieri, Epistole, Ecloge, Questio de situ et forma aque et terre*. Rome: Antenore.

Petrucci, Armando. 2017. Storia e geografia delle culture scritte (dal secolo XI al secolo XVIII). In *Letteratura italiana: una storia attraverso la scrittura* (1988). Rome: Carocci.

Raffa, Guy. 1992. Enigmatic 56's: Cicero's Scipio and Dante's Cacciaguida. *Dante Studies* 110: 121–134.

Roddewig, Marcella. 1984. *Die Göttliche Komödie: Vergleichende Bestandsaufnahme der Commedia-Handschriften*. Stuttgart: Hiersemann.

Rosadi, Giovanni. 1917. *Il canto XXIV dell'*Inferno. Florence: Sansoni.

Saiber, Arielle. 2020. The Textual Nonhumans of Italian Humanism. *California Italian Studies* 10 (1): 1–15.

Seife, Charles. 2000. *Zero: The Biography of a Dangerous Idea*. New York: Penguin.

Shapiro, Marianne. 1990. *Inferno* XXV. *Lectura Dantis Virginiana* 6: 319–331.

Skulsky, Harold. 1981. Thieves and Suicides in the *Inferno*: Metamorphosis as the State of Sin. In *Metamorphosis: The Mind in Exile*, 114–128. Cambridge: Harvard University Press.

Soskice, Janet. 2017. 24: True Desire, True Being, and Truly Being a Poet. In *Vertical Readings in Dante's Comedy*, eds. George Corbett and Heather Webb, vol. 3, 31–50. Cambridge: Open Book.

Swetz, Frank. 1987. *Capitalism and Arithmetic: The New Math of the 15th Century*. Chicago: Open Court Publishing.

Webb, Heather. 2022. *Dante, Artist of Gesture*. Oxford: Oxford University Press.

Zdekauer, Ludovico. 1886. Il giuoco in Italia nei secoli XIII e XIV. *Archivio storico italiano*, ser. IV, 18: 20–74.

CHAPTER 6

Disabled Bodies and Female Souls: From Dante to Marinella

Catherine Bloomer

This chapter addresses both Dante Alighieri's approach to disabled bodies and souls as well as Lucrezia Marinella's (1571–1653) own position within the genealogy of body and soul thought as read in her treatment of gendered bodies in *La nobiltà et l'eccellenza delle donne*, and particularly through her analysis of male bodies as indicative of lesser souls (Allen and Salvatore 2009: 5). The influence of Dante's thought on the body/soul problem, particularly the treatment of bodies with physical differences, such as the disabled, is reread in light of the Renaissance *querelle des femmes* (Malpezzi Price 2008). Marinella is located within this genealogy and analyzed as a reader and interpreter of Dante. As part of her legitimizing literary tactics, Marinella utilizes Dante as both a scientist and as a historian of defective bodies.

C. Bloomer (✉)
Department of Italian Studies, Villanova University, Villanova, PA, USA
e-mail: catherine.bloomer@villanova.edu

M. Pace (ed.), *Dante and the Sciences of the Human*, Palgrave Studies in Medieval and Early Modern Medicine,
https://doi.org/10.1007/978-3-031-69253-6_6

For Dante, defect, particularly physical defect, can be blameless, without sin. Dante signifies human "defect" by several modes—the foremost is defective desire, as exemplified in the poem *Donne ch'avete*; other "defects" include having been born in the wrong place or time (Barolini 1992: 26, 2014: 181–86, 2019: 21–34).[1] In this chapter, I examine the distinct mode which is the physical disability of the body. Dante engages to treat bodily difference separately from evil, or from diseased souls; unlike Marinella, it is not by gender that he separates the two, but rather through the example of physical disability (which he also treats in two distinct modes, one metaphorical, the other literal). I thus primarily analyze physical defect in the *Convivio* and touch briefly on the topic of embryology in *Purgatorio* 25, not only because the *Convivio* is explicit in mentioning disability, but also because it is the more explicitly philosophical of his minor works, and thus I find it to have more organic correspondence with Marinella's *La nobiltà*.[2]

In the *Convivio,* Dante separates the four kinds of defects and impediments that typically prevent a man from gaining knowledge and thus access to Christian thought and God. Dante divides the defects and impediments between body and soul and further between two categories which Barolini calls "interior and exterior" (2022a: 5).[3] Only one interior (and one exterior defect) is deserving of sympathy. Of the interior defects, Dante writes: "Dentro dall'uomo possono essere due difetti e impedi[men]ti: l'uno dalla parte del corpo, l'altro dalla parte dell'anima" (*Conv.* 1.1.3).[4] The two potential—and I stress the level of potentiality in this sentence as these are defects and impediments that can exist, "possono essere," in, seemingly, any and every man—defects within man are divided between the body and the soul, a division that speaks to Dante's conception, here, of the connection, or lack thereof, between body and soul, i.e. the plurality of body and soul. It is a separation necessary for the author to make to clarify his differing attitude toward physical differences versus moral faults. Dante's use of *difetto* and *impedimenti* together emphasizes their double or split application to the body or the soul. He specifies that they are qualifiers of either the body or the soul, and not both. The careful separation between body and soul results in a Dantean division that affects our reading of physical defects. Dante specifies that defects of the body are not related to those of the soul and therefore not deserving of censure.

Dante indicates the clear distinction between body and soul through his treatment of the interior *difetti*: "Dalla parte del corpo è quando le parti sono indebitamente disposte, sì che nulla ricevere può, sì come sono sordi e muti e loro simili. Dalla parte dell'anima è quando la malizia vince in essa, si che si fa seguitatrice di viziose dilettazioni, nelle quali riceve tanto inganno che per quelle ogni cosa tiene a vile" (*Conv.* 1.1.3). Interior defects are split here: one of the soul—the evil desire of malice—and one of the body—sensory afflictions that prevent the human from receiving external information, such as the deaf and "mute" experience.[5] These are both internal defects that affect reception, but Dante qualifies the issue of reception. The deaf body cannot receive anything while the soul with malice in it becomes a follower of vicious delights, by which path the soul receives such *inganno* that everything is vile. The polemic between these defects then becomes simplified: the reception of nothing, which is blameless, versus the reception of evil, which is blameworthy.

For Dante, Adam and Christ are the only human beings created without defect, and he does not mention women as particularly more defective than men (Mazzeo 1957: 717). In the *Convivio*, he lists possible defects that prevent knowledge but again does not specify or single out women (Granacki 2021). Indeed, Dante goes a bit further to insist that those with physical defects are blameless when he writes that they "non sono da vituperare, ma da escusare e di perdono degne; le due altre, avvegna che l'una più, sono degne di biasimo e d'abominazione" (*Conv.* 1.1.4). Dante reserves blame and vituperation for the malice of the soul (as well as the exterior defect of geographic distance or laziness), and not for physical disabilities (and the exterior defect of civic and familial cares).

For Marinella, it is men who are defective. On the title page of her 1601 edition (from which I cite in this chapter), Marinella declares that the second part of her discourse will specifically enumerate, with various historical and modern examples, "che li diffetti de gli Huomini trapassano di gran lunga quelli delle Donne."[6] The first part of Marinella's treatise was written with speed in two months in 1600 as a direct refutation of Giuseppe Passi's *Dei donneschi difetti*(1599) (Schnieders 2014: 193–4; Panizza 1999: 2). Passi's title is explicitly referred to in the second part of her full title, *La nobiltà et l'eccellenza delle donne, co' diffetti e mancamenti de gli huomini*, which seems to be "an emphatic reversal" of his, as Letizia Panizza points out in the introduction to the English abridged translation (1999: 2). But Marinella's focus on defect does not seem to center on the body or the physical; rather, her approach is indicated in her censure of the universality of Passi's own title: she writes that it ought to read "I difetti delle donne malvagie" (154). Her added adjective, *malvagie,* or evil, wicked, is telling; indeed, Panizza calls Marinella's entire approach "a gender-based moral code" (1999: 15). In this way, Marinella expresses a desire to separate two modalities of female bodies: one as tied to sin; the other (and larger whole) as blameless. Largely, she insists it is men who are sinful, women who are not. When Marinella's discourse turns to male defect, she continues to avoid the physical implications of the body, instead, for example, referring to fear in men, "il timore," as a "grandissimo difetto" (305). By focusing on defending sexual difference through attending to the moral superiority of women, Marinella in fact sublimates the reality of the body as a material object. Her focus on virtue and beauty as markers of the female body certainly negates the so-called "defect" of the female (Shapiro 2013: 332–34)—but leaves wholly out of the question—in a purposeful literary and philosophical strategy—the body as a separate object from the soul.[7]

In 1601, Marinella revised and largely expanded the second part of her treatise on the defects of men in which are contained the majority of her citations of Dante (Malpezzi Price 2008). Marinella does not change strategy in the expansion to *La nobiltà et l'eccellenza delle donne*, but rather increases her *exempla*, so as to refute Passi blow-by-blow (Kolsky 2001: 977). Indeed, much of Marinella's methodology is recovered from Passi as a means to refute and discredit his work (Kolsky 2001: 982). She often employs a matching methodology to counter male authors; to respond to Boccaccio, for example, she takes a similarly satirical tone (Allen and Salvatore 2009: 28).[8] Her use of Dante, however, is not combative or reactive, but rather reveals a depth of appreciation for Dante as an authoritative source. Her tone in *La nobiltà* is largely, like Passi's, "vituperative" (Cox 2008: 174), but often she is ironic as well (Cagnolati 2013: 58). Despite the confines or even limitations of her chosen form, her approach to defect cleverly sidesteps the arguments about women as defective men made by Aristotle, Boccaccio, and Passi through her own refusal to write about the (female) physical body without also linking the said body to the soul.

Like Dante, Marinella engages with a hierarchy of defects and establishes whether bodies are to be blamed; throughout her assessment, she uses language that is vituperative. For example, she writes that "il furto quanto sia da ogn'uno biasimato, e vituperato non accade, che io lo racconti percioche è cosa inhumana il volere possedere l'altrui senza alcuna fatica, e quel che è peggio, con la morte di colui, a cui è stato rubato" (295). Despite using similar language, Marinella's rhetorical approach differs from that of Dante's. Marinella simply argues that women are without defect; men are with defect, in both body and soul, and are certainly to be blamed and vituperated. Again, Marinella continuously engages with vituperative language. In responding to male authorities such as Aristotle, she says that "varie furono le cagioni, che spinsero, e sforzarono alcuni uomini sapienti, e dotti a biasmar, e vituperar le donne" (108). She takes issue with these appellations when applied to women. Schnieders cites these pages as examples of how Marinella confronted Aristotle in varying modalities, because Marinella does not construct her refutation based on the content of "sentenze aristoteliche, ma preferisce piuttosto affrontare l'avversario interpretandolo sulla base di un approccio introspettivo" (2014: 198). This introspective approach, I argue, also includes Marinella's wielding of Dante and Dantean language as new authorities in her armament against Aristotle and in her consideration of the body and soul. Dante's relationship to Aristotle differs significantly from Marinella's; he begins the *Convivio* with a paraphrase of the *Metaphysics*, and his discussion of impediments to knowledge adapts the *Nicomachean Ethics*, as I will discuss (Fioravanti 2011: 93–95).

In this chapter, I reconsider the division between body and soul in relation to physical defect, disability, and gender. For Dante, the formation of the physically imbalanced body may carry an associated discourse on the soul—namely, whether it be the disabled body caused by, or a reflection of the morally diseased soul. Both Dantean embryology and the Dantean creation of the shade challenge and complicate the medieval connection of evil with physical defects and sickness (Gragnolati 2005).[9] Certainly in the *Convivio*, Dante's approach allows for the separation of earthly physical disability from vice or virtue of the soul.[10]

The division between body and soul has a long tradition of placing the body first and foremost as a male able-bodied entity. For Dante and the major philosophers before him—Aristotle, Augustine, Aquinas, Bonaventure, among others—the human body is not possible without that which animates it, the soul. A body without a soul is dead; even plants have a (vegetative) soul; even animals have a (sensitive) soul.[11] It was Aquinas' teacher, Albert the Great, who argued that the soul carried and conferred the form of the body.[12] Dante largely follows Albert, who also defined the intellectual soul as the most noble; the better the person, the more intellect shines within them (Callus 1960; Trovato 1990). Dante rejects Averroes' separated soul and does end up closer to the unicity of form, i.e., the idea of the rational soul as the full carrier of the self (Gilson 1967). Yet his treatment of physical defect as separate from defects of the soul inclines more to the pluralists.

At Marinella's time, the classical philosophical language is still the one available to Dante, with the obvious addition of Renaissance thinkers such as Jacopo Zabarella and Francesco Piccolomini, and the context of the *querelle des femmes* (Allen and Salvatore 2009: 11). Marinella's scientific approach points to a contradiction in the genealogy of body and soul thought—it is, as I have stated, a genealogy previously conceived of primarily in male terms. Marinella's work represents the moment when this contradiction comes to life. She is responding to two separate theories of sex difference, resulting in what Emilio De Tommaso (2020) appropriately terms an ontological difference of gender. Marinella's two theories to which she is responding are Platonian and Aristotelian. Allen and Salvatore (2009: 11) call the former "sex unity," and the critics define this Platonian "sex unity" as claiming equality between the sexes; the latter Aristotelian theory is called "sex polarity," in which women are seen to be imperfect men. Allen and Salvatore point out a third theory, gestured toward by Augustine and Anselm, which is exemplified by Hildegard of Bingen, and called "sex complementarity," in which "man and women are equal in dignity and worth, but there are philosophically significant differences between the sexes" (11). Marinella in turn develops "the first completely philosophical rationale for a fourth theory of sex identity, namely *reverse sex polarity*"(11). Her reversal of Aristotle was first suggested by one of her major sources, Henry Cornelius Agrippa, but Marinella goes beyond Agrippa's short discourse (Allen and Salvatore 2009: 12; Malpezzi Price 1994: 108). I propose to read Marinella's theory of sex difference in terms of a Dantean position within the debate on the unity or plurality of the soul. It appears that Marinella, who sees the body as a cipher for the soul, is a staunch proponent of the unity of the soul. Her view of unity results in an application of "reverse sex polarity" to the soul as well as to the body. Thus, when she writes of the body, she is always writing also of the soul (Deslauriers 2019).[13]

Marinella offers a thought-provoking adjustment to both Dante and her contemporaneous academic circles by introducing the question of gender to the development of the soul, using the unity of soul and body to solidify her claims of female superiority. Marinella argues for the superiority of the female soul and the inferiority of the male (Willer 2013). In refuting Ercole Tasso's accusations against women, Marinella writes:

io concede, che quelle cose, che nascono contra lo intento della natura sieno mostri, e vitio, ma ben nego, che le Donne a tal modo nascano prima percioche i mostri rade volte si veggono, e poi sono dalla natura generati: onde per lo contrario veggiamo più Donne, che Huomini nascere, la onde io direi, che gli Huomini siano mostri, generando sempore la natura maggior copia del migliore, & mino quantità del peggiore. Oltre a ciò tanto è inteso dalla natura la generatione della femina, quanto quella del maschio, se ella vuole eternare la spetie de gli Huomini, ricercandosi alla generatione la Femina, & il maschio. (165)

Marinella espouses here a philosophy of difference that denies women as monstrous and instead casts men in that role. She claims men to be monstrous based on nature producing more women than men, as women are needed for the generation of the species. Monstrosity has long been linked to the discourse of disability (Godden and Mittman 2019), while Marinella applies that discourse to men, thereby arguing that it is *male* bodily difference that is representative of defect, monstrosity, and lack. She refers to women simply as "migliore" and men as "peggiore," thus creating a binary between the better and the worse, the body that generates and adorns, and the body that is monstrous. Thus, when she writes of defect, she writes of evil.

While evading the old idea of the female body as defective or monstrous, Marinella turns to ideas of physical deformity as constitutional and symbolic, which is reflective of the state of the soul, and to how a certain kind of (male) body is an index to a defective soul. Her third chapter, *Della natura, & essenza del Donnesco sesso,* opens with a citation of Moderata Fonte in defense of women's nobility, with which Marinella takes issue saying that "io direi che l'anime delle donne fossero nella lor produttione vie più nobili di quelle de gli huomini; si come da gli effetti, e dalla bellezza del corpo si può vedere" (17). Fonte's equalizing is not enough; indeed, Marinella insists on female superiority in their souls, on the basis of the beauty of their bodies. She thus reads the body as a cipher for the soul. As De Tommaso (2020: 65) explicates:

Marinella ritiene che non si debba ascrivere la differenza tra donna e uomo alla loro fisiologia, come di fatto sottende l'ipotesi del *mas occasionatus*, ma che essa sia ontologica, relativa cioè alle idee presenti nella mente di Dio, sulla base delle quali è stato creato l'universo intero. Si tratta, secondo la pensatrice veneziana, di una differenza non meramente materiale, perché uomini e donne non sono diversi semplicemente nei corpi, ma lo sono soprattutto nelle anime.

Marinella links the body to the soul and thus to its endowment by God. Marinella's discussion of physiological difference plays out on the intellectual capabilities of women as well. She indicates firstly that intellect cannot be measured in quantifiable terms; thus, men cannot have greater intellect than women. And thereby dismantling men's biologically superior intellect, she simultaneously dismantles the "justification of women's subordination to male authority as a natural consequence of their biologically predetermined inferior intellect" (Sinclair 2020: 40). This question of intellect leads to the question of the soul, in which, for Marinella, women will be again superior. She ties women's superiority in these cases to the inferiority of men's appetites, heat, and thus reason, which all contribute to making men "intellectually inferior" (Deslauriers 2019: 726). She espouses the medieval connection of evil with ugliness, physical defects—both of which Dante is attentive to separate—or above all other cases, maleness, in her efforts to elevate and link beauty, haleness, and the feminine. For Dante, intellect is also created or endowed by God; this intellect is breathed in by God to every human embryo; it would be unthinkable to consider God makes an error in this act, thus the defect must be contained within the body, formed by the sperm or blood of the father (Gragnolati 2003).

Marinella closes her second chapter with a simple statement that woman is made from man's rib, and hence superior as man is made from mud—a popular point that Arcangela Tarabotti phrases similarly in her *La semplicità ingannata* (1654)—while Marinella's original 1600 edition added that woman was thus created in Paradise (Willer 2013: 210). Marinella clearly states that man is created first in order to form women, and this addition of hers is distinct and original, not present in Agrippa, one of her sources (211). Not only are male and female bodies different, the male body was *only* created to form women. Furthermore, this difference "si ritrov[a] già nella mente di Dio," and their physical difference is also reflected in their souls. As De Tommaso points out, "La prima interessante implicazione di questa visione è che la distanza tra maschile e femminile non possa ridursi alla mera fisiologia, ma si connoti come una costitutiva differenza ontologica, di cui l'aspetto esteriore segna la misura" (2020: 73). Marinella thus destroys the male as a model for humanity and elevates the woman "come piena manifestazione dell'umano" (73).[14] Marinella reads gender difference not as an accident or even an accident of nature, but rather as an indication, made by God, of women's superiority. Men's defects are then read as both reflections of their bodies and their lesser souls.

I examine this de-privileging of the male by paying particular attention to Marinella's citations of Dante and by considering his own scientific approach to the body and soul, his consideration of bodies that are not able-bodied. Dantean citations, particularly the examples of men's perfidy, are utilized by Marinella to her advantage in her defense. In fact, Marinella's full capability as a reader of Dante shines in the second part of her treatise, as she embarks on a quite literal reading of Dante that sees her cite examples of historical men in hell as evidence of men's defect. The richness of Marinella's poetic culture has been investigated by Adriana Chemello who writes that "Marinella's treatise can be seen as the exhibition of a 'library' opened both to show itself off and to illustrate the intellectual qualities and profound erudition of the author" (2000: 473). Not only do her sources reveal a popular culture of authority, but they also reveal her own preferences.[15] While Passi's sources are largely classical or biblical, Marinella's are in large part vernacular, including the *tre corone* (Baradel and Zancan 1983: 162–66).

In particular, Marinella had access to both her father's—who admired Aristotle—and brother's libraries (Allen and Salvatore 2009: 7; Cox 2011: 240). Both were doctors, as was her husband and Passi himself (Ray 2015: 93). Marinella even dedicated her work to another doctor, a friend of her father, Lucio Scarano (Ross 2009: 198). Her father's manual on women's health, *Le medicine partenenti alle infirmità delle donne* (1563; revised and expanded, 1574), dedicated to women, focuses on the safeguarding of marriage as an institution to procreate but avoids the overtly misogynistic language that, for example, Boccaccio employs in linking menses to defects.[16] While Marinella's treatise is aimed at Passi, she also takes on Boccaccio, Speroni, Namur, and Aristotle, among others.[17] In refuting Boccaccio, Marinella writes that "le Donne non sono animali imperfetti" (177), which is perhaps also an intended blow against Aristotelian thought (Deslauriers 2017: 2528).[18] Marinella writes specifically against the Aristotelian idea of women as defective, but every counterpoint of Marinella's refutation is part of an ever-expanding web that seeks to confute many of the philosophers before her.

Dante's *Convivio* begins with a paraphrase of Aristotle's *Metaphysics:* that all men desire to know. Dante then looks at obstacles to the state of perfection, or knowledge (of God). His treatment of *malizia,* or the defect of the soul, owes much to the first chapter of the seventh book of the *Nicomachean Ethics* (Fioravanti 2011: 93–95). The defects and impediments of the body, however, deviate from the *exempla* given by Aristotle, who specifies as blameworthy blindness that is acquired through our own action, from drunkenness or some other form of self-indulgence in contrast to blindness from birth (blameless). Dante changes the focus on the voluntary act and refocuses the discourse on the nature of the defect; he also omits the example of the blind man and inserts instead the example of the deaf, "mute," and their similar, thus potentially expanding the category of disability. Dante puts defects into a complex hierarchy: blameworthy is the evil soul and laziness (defect of place); blameless is the body that cannot receive information (i.e., the deaf) and the distraction of civic cares. Dante's is a nuanced take on defect that considers whether the presence of the condition, without any qualifiers of how it was endowed, can be blameless.

Unlike Dante, Marinella does at times reflect on the physicality of the female body, or as Deslauriers puts it, the "physiology of sex" (2017: 2526). She certainly ruminates on the physiological difference of temperature, following Aristotle in treating heat as a moral and intellectual indicator; yet, she claims that female bodies are temperate, and their blood too.[19] Marsilio Ficino is also an influence on this discourse, and Marinella utilizes his medical theory in considering temperature, or heat, as an indicator of the soul.[20] This physiological difference, then, is again directly linked as an indicator of the soul's nature. Marinella's reappropriation of Aristotle's biology—especially what he states in *De partium animalium* 2.4, 650b 33-651a 1—makes the hotter men thusly angrier and less noble (Deslauriers 2017: 2535; 2019: 727–28). The discussion of temperature is medical in nature, an inverted model of Galenic humors (Ross 2009: 286). Marinella's thought on the science of her day, as well as her rethinking of a Dantean system of defect, also reveals her thinking on *auctoritas* (Jordan 1996: 56). Indeed, the argument that moderate warmth contains greater nobility is also found (perhaps coincidentally) in Hildegard of Bingen (Allen and Salvatore 2009: 22). The discussion of natural defects also links to Dante's intercession on behalf of those who were not born noble. The medieval debate spans whether virtues (and vice) are endowed (whether by Nature or by God) or acquired. For

Dante, as Trovato writes (1990), nobility is equated to the rational soul, and is not a stable or immobile quality; nobility, then, is that which is not bad and can be present in all men to varying levels. Dante thus makes nobility a fluid category.

One might trace here Dante's own medical context of "sangue perfetto" in *Purgatorio* 25 as the generative substance, the superfluous blood that can be traced to Aristotle, Galen, Avicenna, and Albert the Great—although it is unlikely that Dante knew directly of Galen. For Dante, science is knowledge. His first guide is reason. In *Purgatorio* 25, when Statius promises to reveal the *vero assoluto*, he engages into a dialogue ostensibly on embryology but that also reveals Dante's scientific knowledge, his opinion of the unity/plurality debate, and the faculty of the rational soul. Statius begins answering Dante's question about the thinness of the afterlife bodies by describing first the development of the body on earth. The process of insemination is described as blood moving to the genitals, becoming semen, and solidifying the mother's passive blood; for Dante, this power becomes the soul, in a series of evolutionary steps from plant to sponge to animal. This particular phase, in which the organs develop, seemingly indicates the body is formed without yet a human soul (rather with the underdeveloped animal soul that had been a vegetative soul). Then it lives (vegetatal), feels (sensitive/animal), knows (intellect), reflects (self-awareness). Cerbo (2001: 29) draws the interesting connection that "[i]l fenomeno generativo è impostato fisicamente, anzi fisiologicamente, alla stessa maniera in cui Guido Cavalcanti tratta il processo dell'innamoramento." Ureni (2010), who is working on the ethical side of Galenic humors and imbalance in conjunction with the medieval ideal of perfection that does not exist on earth, writes on the formation of the shade in the afterlife. She notes the Dantean creation of the shade operates by the same faculty as the creation of the physical body. Her conclusions can offer significant implications for the creation of the (disabled) body and the connection of evil with physical sickness, as in the Galenic principle of cure by *contraria* would propose disease as a result of excess or deficiency, and afflictions of the soul that present as anger, fever, etc., as treatable through the body. It is this anger to which Marinella refers, and the connection between moral and physical afflictions is directly related to the physically imbalanced body. It bears repeating that Dante's science of the human, at least in *Purgatorio* 25, is not one explicitly conceptualized in terms of gender difference, as instead Marinalla's is restructured.

In another instance in which Marinella treats physical difference, she responds to Ercole Tasso and seems to gesture towards the physicality at least of male bodies: why shouldn't "la Donna [lamentarsi] del marito; percioché ella sia più giovane, più ricca, più nobile, più savia, e più di lui fornita di bellezze," she asks (167)? Why should young women content themselves with "un Vecchio isdentato con gli occhi, & col naso gocciolante?" (167). It seems to be no accident that physicality enters the discussion of marriage. She shifts from a moral to an "acerbic" register as she outlines that men are bad examples of conduct and turn into animals (Sinclair 2020: 437). In the twenty-fifth chapter of the second part, *De gli huomini giuocatori,* which turns to gamblers, Marinella also laments about men's deformities. She deplores that not only are these men who gamble, but that often they are old, decrepit, and also "paralitici, con gli occhi scarpellini, che non havrebbono veduto uno Elefante in uno campo di neve, con dua paia di occhiali al naso, mettere al punto, poiché non potevano altrimenti giuocare. Altri pur vecchi, e infermi; perché non possono muovere le mani, fanno, che alcuno altro giuochi per loro" (364). Virginia Cox takes note of this example as well, citing it as indicative of Marinella's "sarcastic wit" (2011: 240). Marinella makes the male defective, monstrous, comic; here they are paralyzed, unable to see even an elephant, and old and infirm. Her approach in this example indicates her larger refutation of the Aristotelian idea, that Passi espouses (1599: 23), of women as a mutilated or erroneous man. She is again taking up the strategy that Allen and Salvatore call "reverse sex polarity" (2009: 11). In addition to making the male monstrous, Marinella employs various strategies to address and combat the Aristotelian idea of women as defective; among these, as I have stated, she turns to female beauty and intellect as exemplars of female superiority. Marinella may appear to treat some women's bodies as exemplary cases, such as the physical and moral strength of the Amazon, as Sandra Plastina considers (2017: 497–502). However, in challenge to the idea of an exceptional woman warrior, Marinella indicates that were women trained in the same labor as men, they would be just as strong (Allen and Salvatore 2009: 25). Indeed, for Marinella strength is not indicative of superiority; it does not confer *scienza*, or knowledge; nor does it confer justice (Jordan 1996: 64). Dante's own concept of *scienza* may offer relevance to Marinella's discourse. In the *Convivio,* he argues that the reason that men desire (*desiderano)* knowledge is that everything, coming from the "prima natura," from God, is to incline toward their own perfection.

Thus, *scienza* is the ultimate perfection of our souls, in which is found our ultimate happiness, a desire to which we are all subjected (Barolini 1992: 99–101). While Dante then discusses the defects and impediments that can exist in every man and which prevent men from gaining *scienza*, Marinella seems to make such *scienza* a quality only achievable by women.

Marinella's first two citations of Dante are utilized to support her analysis of *donna*, the first of five names for woman, and she thus employs Dante as evidence to support an Aristotelian idea that names (or rather proper nouns) uncover an essential component of things.[21] Marinella's discussion of *donna* is meant to refute Passi's own opening chapter of *Dei donneschi difetti* in which he opines on the etymologies of *femina* and *mulier*. Marinella employs both Dante and Petrarch in her analysis, and indeed, the Dantean citations immediately follow citations of Petrarch, initially establishing the pair as linked authorities on the Italian language. She cites the two crowns as examples of the masculinization of the noun *donna*, first from the *Rvf.* 360.65: "Per inganni, e per forza è fatto Donno" (6) [Per inganni et per forza è fatto donno], and then immediately from *Inf.* 22.83: "Ch'hebbe i nemici del suo Donno in mano" (6) ["ch'ebbe i nemici di suo donno in mano"]. Her citation of Dante identifies an etymology for her own terminology for women and for defect. She then turns to examples of verbs and adverbs, citing the *tre corone*, with Dante's *Par.* 7.13 last: "Per quella riverentia, che s'indonna" (7) ["Ma quella reverenza ch s'indonna"]. She concludes that this *donna* "denota signoria, e imperio: ma placido dominio a punto corrispondente alla natura della Dominante" (6). These initial citations establish Marinella's use of Dante as a clear authority in matters of language. And still, her usage of Dante is at this point indistinguishable from her citation of Petrarch. However, her opinion of Dante (and Petrarch) is not one of antagonism, as it is for Boccaccio and the others against whom she directs her discourse.

In this discussion of female names, Marinella includes a fourth out of the five names for woman, *Ischia*, which, like *Eva*, was not considered by Passi (Chemello 2000: 467). Marinella writes that:

> questo nome dimostra un fuoco celeste, divino, e incorruttibile, la cui natura è di perfettionare l'anima ne' nostri corpi chiusa, di eccitarla, illustrarla, e in somma renderla partecipe di divina perfettione, allontanandola da ogni bruttezza terrena. Si vede risplendere questo celeste fuoco nella bellezza del corpo del sesso donnesco, come al suo luogo proveremo. Che

> si può dire di questo nome, se non che sì come le celesti cose sono più nobili delle terrene, così che questo superi di gran lunga tutti gli altri, già che gli huomini rende partecipe di divina essenza? Onde si può ben chiamare infelice quell'huomo, che si trova haver priva la casa d'un tal fuoco, che lo ecciti e svegli a contemplare il Cielo. (11)

As Allen and Salvatore argue (2009: 18), Marinella works here with a Platonic idea of the soul. Her reading of women's souls as containing divine light is also certainly indebted to Petrarch (Chemello 2000: 469). She is in conversation with Plato, through Ficino and Plotinus, or even perhaps through Dante's ideas of divine light. Her discussion of this fourth name mirrors the course of the third chapter, which turns from poets to philosophers. Marinella takes what is visible to be the clearest proof of gender difference on the part of the soul, and thus once again takes the body to be a clear indication of the state of the essential being, or the soul (De Tommaso 2020: 70–71).

The distinction between Dante's approach to defect and difference and Marinella's universalizing or totalizing gender-based approach to defect is particularly clear in her third citation of Dante in her text, which appears at the beginning of the second chapter, *Delle cause, dalle quali dipendono le Donne*. Marinella cites Dante as evidence that, though God is the prime creator, his works may have differing values. She notes that "Dante volendo dimostrare la diversità degli effetti della somma bontà disse nel suo Paradiso: 'La gloria di colui, che'l tutto move / Per l'universo penetra, e risplende / In una parte più, e meno altrove'" (13–14) ["La gloria di colui che tutto move / per l'universo penetra, e risplende / in una parte più e meno altrove," *Par.* 1.1–3]. Marinella cites these verses as evidence of why a woman is more noble than a man, despite having the same creator. Dante's *terzina*, in fact, refers to what Barolini has called "the paradox of *più e meno*"—it stages the idea of difference and of multiplicity that can exist and indeed coexist with unicity (1992: 166). God's light and glory penetrates the universe—but that it shines less in some places is indeed an issue of reception, and it is this idea of reception that plagues Dante in *Convivio* with regard to the deaf. He seems genuinely concerned about who can or cannot receive the light or glory of God in physical, and scientific terms, and whether humans are responsible for said ability or inability. Marinella instead applies these lines universally to men and indicates also that she finds them as a group entirely responsible for their condition—whereas Dante, perhaps more nuanced in his approach, finds some physical defects to be blameless.

For Dante, the human body that is imperfect is universal, and not the fault of God, nor of the individual. Indeed, Marinella refers to this idea of imperfection in the second part, her fifth chapter *De' Superbi, & Arroganti*, where she admonishes: "Non v'accorgete voi, che sete vermi, come narra Dante" (227). Marinella is referring to Dante's lines in *Purgatorio* 10—"non v'accorgete voi che noi siam vermi / nati a formar l'angelica farfalla, / che vola a la giustizia sanza schermi?" (lines 123–26). Marinella switches the pronouns, making Dante's inclusive and universal *noi* into a directed *voi* that targets males. Again, she takes what is universal for Dante and applies it wholly to the male. This clever switch also clearly indicates the high—and feminist—level at which Marinella is operating as a reader of Dante. She is not misreading, but actively reinterpreting. The switch also seems to call attention to a perceived male readership of Dante; from Marinella's perspective, she is not included in Dante's *noi*.

Following a citation of Petrarch (Baradel and Zancan 1983: 158), Marinella also uses a supposed *ballata* of Dante's—whose author is actually considered to be Cino da Pistoia—to indicate not only women's superior beauty, but also women's superior origins: "Et Dante in una sua ballata dice, che guardando il viso a Madonna diverrà beato a guise d'Angelo: 'Poi che satiar non posso gli occhi miei / Di guardare a Madonna il suo bel vi[so]'" (27–28). Marinella's reading of this *ballata* as Dante's may be forgiven doubly; first it would have been attributed to Dante at the time; secondly, Dante's influence on Cino da Pistoia, who is now considered its author, is undeniable (Giallongo 1995: 193). Nevertheless, we may read Marinella's "Dantean" ballata as further evidence of her use of the poet's words as a rhetorical device in her own work. Love is ennobling, beatifying, in *Poi che saziar non posso gli occhi miei*. Marinella reads the poem in line with Petrarch's praise of Madonna Laura, signifying that female beauty leads to God—a supremely (neo)Platonic idea, to be sure, and an inventive use of vernacular poetry. One might wish to see a similar treatment of Dante's Beatrice, but Marinella does not go there. Rather, she cites more vernacular poets (Annibale Caro, Bernardo Tasso, Giovanni Guidiccioni, Petrarch) before offering her own version of Diotima's ladder, where Diotima herself is listed by Marinella as one of her "donne scientiate" (51). Here Marinella changes the ladder to be "di anello in anello" a "dorata catena" (30).[22] Lisa Shapiro calls this "shift in metaphor" (2013: 332) a clear indicator of Marinella's adaptation of the

Platonic staircase, moving from independent steps to a linked process. Marinella's first ring is the (female) body, the second is the soul, with chains or links leading ever upwards to God. And thus she exclaimes: "che dono, oh che doti, oh che maggioranze sono queste delle donne: poi che con la lor bellezza può alzare le menti de gli huomini a Dio" (30–31). Marinella phrases the gift of women's bodies, as part of this link upwards to God, as a dowry, in a material, quotidian, and female term (and refers back perhaps to Moderata Fonte's own reading of *donna* as a *dono*). One may consider whether Marinella includes women in the *huomini* whose minds are led upwards, or whether she in fact reduces the female here to act as an inspiration point. She may be expanding on Marsilio Ficino's idea of beauty as a divine gift, especially his translation of Plotinus' *Enneads*, which follows Plato's *Symposium* closely (Deslauriers 2019: 725). She also cites Plotinus and Ficino to illustrate further this point: "che il corpo delle donne sia più nobile, e più degno di quello de' maschi ce lo dimostra la delicatezza, e la propria complessione, o temperate natura sua, e la bellezza: anchor che la bellezza sia una gratia, o splendore resultante dall'anima, e dal corpo; percioché la beltà senza dubbio è un raggio, e un lume dell'anima, che informa quel corpo" (19). It is Ficino who speaks of transcendent beauty, or of women's body and soul's beauty as a divine ray or gift (Weiss 2022).

Adding to her growing discourse, from the cited *ballata* to a Diotima-like chain, Marinella goes on, first citing Platonists, and later Petrarch. She then states, for example, that the idea of "una leggiadrissima Ninfa" is better than "un rustico, e difforme Satiro" (15). She uses examples of bodies, beautiful or deformed, as indicators of ideals, of states of being. De Tommaso (2020) analyzes this same passage, writing that "Questo implica che, se in natura si esprime una differenza tra donna e uomo, è solo perché essa è già preesistente nella mente di Dio. Dunque, la differenza di genere ha un suo preciso statuto ontologico, che è sancito dalle diverse idee di 'uomo' e 'donna' eternamente presenti nell'intelletto divino" (67). Again, Marinella takes the binary between man and woman to be not only reflected in the bodies and souls but fundamentally formed in the mind of God (Allen and Salvatore 2009: 19).

Returning to Dantean citations, particularly the examples of men's perfidy, Marinella continues to utilize the *Commedia*, revealing her full capabilities as a reader of the text. For example, she uses Dante's lines to correct an error of Passi's own citation of Dante; moreover, she uses

Dante to define the concept of fraud, citing from *Inferno* 32 to illustrate hypocrisy in her own thirty-second chapter, *De gli Hippocriti, & Santoni*. This correspondence suggests Marinella's evident knowledge of Dante's *canti*. Marinella of course also corresponds to Passi's chapters, from whose chapters her second part takes its shape, but I add that Passi's limited citations of Dante do not feature a similar correspondence to Dante's themes (Schnieders 2014: 194).

In the following direct address to Passi, Marinella uses Dante to get the last word. In her twenty-seventh chapter, *De gli huomini loquaci, & cicaloni*, Marinella states that these men "credono bene spesso [...] di levarsi con molte parole i molti difetti" (373). She then pointedly corrects an "Error del Passi" by writing that she would like "che Gioseppe Passi sapesse, che Dante pone nello'nferno la schiera de' cianciatori, e non delle Donne cianciatrici; e però non ista bene a volere raccontare un discorso delle Donne parlatrici, ma porvi la schiera de gli huomini cianciatori in vece loro" (376). She is certainly responding to Passi's use of Dante in his argument against loquacious women, where he writes that "Dante nel suo Inferno pone i loquaci da varie colpi di spada tagliati, e divisi dal Demonio, dicendo; 'Un diavolo è qua dentro che n'accisma / Sì crudelmente al taglio della spada, / Rimettendo ciascun di questa risma'" (287) ["Un diavolo è qua dietro che n'accisma / sì crudelmente, al taglio de la spada / rimettendo ciascun di questa risma] (*Inf.* 28.37–39). The verses Passi cites are from *Inferno* 28 and indeed come from the infamous speech of Muhammad, who is a man, while the *canto* contains no description of female souls (Marinella herself follows this trend as the second half of her work contains only male examples from Dante's oeuvre). Marinella thus displays herself to be a far more astute and knowledgeable reader of Dante than Passi.[23]

Marinella refers to Dante's fraudulent speakers, and she later invokes specific examples of the fraudulent in her eleventh chapter, *De gli huomini Fraudolenti, Traditori, Perfidi, & Spergiuri*. Marinella cites again from *Inferno* and following a citation of Ariosto on fraud, when she says that Dante "in questo modo haudendola veduta nello Inferno" (272). She cites the following Dantean lines: "E quella sozza imagine di froda / ... / né fuor tai tele per Aragne imposte" (*Inf*. 17.7–18). Recognizing she has cited a rather long Dantean passage, she then abruptly states "Ma basti fin qui di haver narrato quello, che tiene in sé, e fuori di sé la fraude, et veniamo a gli essempi..." (273). Marinella shows her hand—she enjoys displaying her many examples and would rather leave off defining the sins. She seems to accept Dante as the moral authority and takes on her own role as a cataloguer of *exempla* that neatly corresponds to his own moral order.

In another example, she uses Dante to aid her definition of a defect. In the thirty-second chapter, *De gli Hippocriti, & Santoni,* immediately after defining hypocrisy, Marinella writes: "Però Dante per mostrar, che costoro non hanno in loro alcuna cosa di vero, eccetto la superficie, e il di fuori, li chiamò gente dipinta nel suo inferno, dicendo, 'La giù trovammo una gente dipinta, / che giva intorno assai con lenti passi / piangendo.'" [*Inf*. 23.58–60] (398–99). Here, Marinella cites the introduction of the hypocrites from *Inferno* 23, in the second half of which the pilgrim finds their *bolgia*. Marinella is struck by Dante's language, and she is making a gloss of the "gente dipinta"—indeed, she is thus making herself a commentator of Dante. She continues that "Gli huomini di tal fatta paiono essere tutti pieni di carità, e di devotione, e mostrano non si movere mai, se non per zelo di Dio, e utile del prossimo. Fanno questo per essere tenuti buoni, e bramano di parere buoni, per non essere tenuti Hipocriti, perché sanno bene quanto questi buoni huomini sieno odiati, e abhorriti dal mondo" (399). She has captured the sense of Dante's hypocrites, using his *terzine* as the basis for her own definition. This usage of Dante would harken back to her utilization of his lines in defining or etymologizing *donna*, but also expands her role as she now comments on the significance of those episodes.

Marinella's use of Dante's poem now turns to take his shades, and their punishments, as literal historical evidence of defective or evil men.[24] She goes beyond Passi in this usage; Passi discusses, for example, Mirra and Megera, citing Dante, but aside from these two mythological figures, he never uses Dantean examples of contemporaneous or historical women as defective (1599: 31, 42). One of Dante's mythological figures, Cacus, is cited by Marinella in the sixteenth chapter *De gli huomini ladri, assassini, corsali, & rapaci*. Cacus is described as the son of Vulcan and "il primo ladrone d'Italia" (296). She relates two accounts of Cacus, one from Vergil's *Aeneid*, and one from Dante's *Commedia*. Marinella does not focus on Cacus's body or physicality, and her text shows little indication that Cacus is a centaur and not a man. Here she again reads Dante as a moral authority, whereby his system of punishment is treated by Marinella as an absolute indictment of the mythological figure. Indeed, she repeats the emphasis on Dante having seen him in hell—"Dante lo vide nello inferno" (296)—and she takes Dante's journey, or his account, nearly as the absolute fact of an eyewitness.

She immediately returns to the examples and vivid specificity of Dante's male historical figures, but she treats Dante's shades and their descriptions as real men who have been condemned by a true authority. Marinella describes Vanni Fucci as a "ladro di cose sacre" (297), and she adds lines from *Inferno* 24 to further cinch her condemnation. Indeed, what Cox calls the "mischievous ingenuity" of her *exempla* (2011: 240), is where her citations of Dante multiply. Marinella embarks on a quite literal reading of Dante that sees her cite examples of historical men in hell as evidence of men's defect.

In her nineteenth chapter, *De gli huomini Incantatori, Magi, & indouini*, Marinella turns to Michael Scott as a "gran Mago" (326). She writes that he "senza preparatione di cosa alcuna, convitava molte persone a mangiar seco, e quando era l'hora del desinare costringeva i demoni, dei quali era tutto suo, a condurli cibi di diverse parti, onde di lui ragionando, dice Dante: 'Quell'altro, che nei fianchi è cosi poco, / Michele Scotto fu, che veramente / de le magiche frodi seppe il gioco'" (326–27) ["Quell'altro che ne' fianchi è così poco, / Michele Scotto fu, che veramente / de le magiche frode seppe'l gioco." (*Inf.* 20.115–17)]. She later adds another Dantean example that immediately follows Michael Scot: "Asdente Parmegiano calzolaio, huomo grosso, e idiota si diede all'arte dell'indovinare, e però Dante lo pone nell'inferno, e dice – 'io vidi Asdente / c'haver atteso al cuoio, e a lo spago / Hora vorrebbe, ma tardi si pente'" (333) ["Vedi Guido Bonatti; vedi Asdente, / ch'avere inteso al cuoio e a lo spago / ora vorrebbe, ma tardi si pente" (*Inf.* 20.118–20)]. Marinella seems to indict Asdente equally for his physical qualities, being fat, for his lack of education or mental defect, "idiota," as well as for Dante's reasoning, i.e. practicing the art of "indovinare." Her inclusion of Asdente immediately after Michael Scot also indicates she may be working from a text of the *Commedia* that is open before her. These are perhaps the most unique citations of Dante in which Marinella distinguishes him as a useful source: she now neatly does away with any mythological and magical figures, and instead turns her focus directly to historical men, taking Dante as an authoritative source.

Marinella's is a very particular reading of Dante, and it is part of a formula that Marinella repeats throughout the second half of her discourse. She treats Dante's sinners and punishments as a very literal moral system, and she takes Dante nearly as a historian, using his verses now to corroborate further examples of bad men: Marcus Licinius Crassus, Ciacco, and Ruggieri degli Ubaldini and Ugolino della Gherardesca—the last of whom she actually exonerates.

Marinella also evokes the vivid specificity of Dante's male historical figures. In her first chapter (*De gli huomini avari, & desiderosi di denari*) of the second part, Marinella begins a discourse again avaricious men, listing the Roman example of Marcus Licinius Crassus, against whom, as Marinella notes, Cicero writes. In this regard, she cites verses of Petrarch on Crassus as well.[25] To further establish his guilt, she cites from *Purgatorio* 20: "Et Dante; dice a Crasso: 'Dici che fai, di che sapor'è l'oro?'" (188) ["Crasso, / dilci, ch'l sai: di che sapore è l'oro?" (*Purg.* 20.116–17)]. Notably, Dante is recalling instances of the avaricious as part of the *exempla* of the purgatorial terrace, as he is not actually speaking to Crassus face-to-face. But Marinella seems to really have it in for Crassus, against whom she again inveighs as well in the second chapter against *Invidia* (202). Again, Marinella is treating Dante's citations of men very literally. She takes the approach not only of examining contradictions (especially Passi's), but also of what Constance Jordan calls "historical relativism" (1996: 57–61). Marinella had also turned to history in her fifth chapter on learned women, seemingly indicating a shift in her approach from Aristotelian and Platonic to what Allen and Salvatore call "a cohesive, intellectual framework" (2009: 23). This historic framework sees Marinella cite Dante particularly with a focus on his Italian historical men.

In the third chapter, *De gli incontinenti, cioè Golosi, ubriachi, & sfrenati,* Marinella adds another Dantean character to her growing and impressive list of *exempla*: "Che dirò di Ciacco, che in lingua Fiorentina vuol dir porco, parlando il Boccaccio di lui dice. Essendo uno in Firenze da tutti chiamato Ciacco, huomo giottissimo quanto alcuno altro fosse giamai, e con quello, che segue. Dante lo pone nell'inferno, e a lui fa dire cosi: 'Voi Cittadini mi chiamaste Ciacco / Per la dannosa colpa de la gola, / Come tu vedi a la pioggia mi fiacco'" (210) ["Voi cittadini mi chiamaste Ciacco: / per la dannosa colpa de la gola, / come tu vedi, a la pioggia mi fiacco" (*Inf.* 6.52–54)]. Ciacco indeed figures among the *golosi* of Dante; again, Marinella seems to order her Dantean examples as direct references to her chapter titles (which of course correspond or respond to Passi's own). The interest of her addition of Ciacco, whom she takes as another historical figure, for Dante scholarship, which still debates both his identity and the idea of whether his name signifies 'pig' (Marchesi

1999), can also be connected to disability. Marinella takes the male to be a defective animal, which in turn constitutes another example of her reversal of the Aristotelian biological framework. Marinella moves immediately from this example to that of the companions of Odysseus eating the cattle of the sun, which she uses almost as a metaphor for Dante's own divine punishment, which she glosses as such: "Cioè il fuoco della gola col fuoco celeste" (210). The comparison calls to mind Dante's own *contrapassi* and their neat correspondence to each shades' sin, as in *Inferno* 28, where Bertran de Born describes his own punishment, carrying his decapitated head like a lantern, using the word itself. *Contrapasso* is a dark twist of metamorphosis, or of the desire for a body not one's own. Dante's use of *contrapasso* highlights a universality of corruption made visible on an aerial body—and to write about disability in the *Commedia*, one might be tempted to write about *contrapasso* as though it were directly representative of the earthly afflictions that affect the physical body. But these are punishments that affect shades, aerial bodies, not human flesh. Marinella's interest continues to remain on Dantean figures that are historical, that is, who once possessed real, male, bodies, and she continues her almost frenetic citations.

In the middle of her long tenth chapter *De gli huomini crudeli ingiusti, & micidiali* (250–71), Marinella inserts a long citation of Ugolino from *Inferno* 33. She writes: "Non voglio già lasciar l'Arcivescovo Rugieri, il quale fece morire di fame il Conte Ugolino, ma voglio lasciar l'istoria, e mettere i versi del nostro Dante, il quale fa così dire al Conte Ugolino nel canto 33 dell'Inferno" (258). Marinella indicts Ruggieri, not Ugolino. Further, she here states her own desire as author—she does not yet wish to move on from her *exemplum* of Ruggieri, but she does wish to change modalities: she is leaving history and supplanting it with the lines of Dante. This is the clearest statement of her intention in citing Dante, whereby she uses his verses as greater corroboration of Ruggieri's evil. Effectively, her use of Dante is a new, revisionist history. She first cites the *terzina*: "Che per l'effetto, de' suoi mal pensieri, / Fidandomi di lui, io fossi preso, / E poscia morto dir non è mestiere" (258) ["Che per l'effetto de' suo' mai pensieri, / fidandomi di lui, io fossi preso / e poscia morto, dir non è mestieri" (*Inf.* 33.16–18)]. In these lines, Ugolino explains why he is near Ruggieri and how his trust in him was betrayed; perhaps the focus on this citation explains in part why it is Ruggieri whom Marinella condemns, and not Ugolino. Marinella skips over the subsequent *terzina* because she does not wish to go on too long: "E più sotto dice, per non essere troppo lunga, la qual cosa non mi piace" (258). This is a subtle acknowledgment of the space she gives to Dante here, which is in fact rather long. She then cites the long passage that imagines Ugolino's ordeal: "Già eran desti, e l'ora s'appressava / … / Poscia, più che'l dolor, poté'l digiuno'" (*Inf.* 33.43–75). She continues her reading of the *canto*, indicating that she is making her way down through the verses: "E più sotto esclamando Dante, mosso a misericordia di tanta impietà dice…" (259). Marinella is referring here to the materiality of her experience as a reader of Dante, once again giving the sense that she is reading from an open text as she writes. She also indicates her own sense of Dante's passion, here citing the exclamation against Pisa from *Inf.* 33.79–87 that follows the episode of Ugolino. Marinella is trying to abridge Dante, to offer a synthesis of his episode and encounter with Ugolino that serves her own argument. Her greatest concern lies with the innocent children of Ugolino's. Her pity is, perhaps, not entirely for him, whom she describes as "il misero Conte," but rather more so for the children, the "quattro figliuolini innocenti" (259). But it is Ruggieri who is "il crudelissimo huomo" (259). Marinella in part exonerates Ugolino—whom we must remember Dante himself places with Ruggieri among the betrayers. This is thus an episode in which her own sensibilities reread Dante's punishments. She is making Dante's verses both replace and overwrite history—and at the same time, her interpretation of Dante overwrites his own condemnation of Ugolino. Although she is rewriting Dante's own representation, Marinella's treatment of Ugolino comes closer to a more nuanced and thus Dantean approach to sin, defect, and the poetics of exoneration.

Adding to these examples of Italian men, Marinella writes of Ciampolo da Navarra: "Dante fa mentione di quel barattiere nato in Navarra, nel Canto 22 dell'Inferno, il quale risponde a Virgilio, dicendo 'Io fui del Regno di Navarra nato, / Poi fu famiglio del buon Re Tebaldo, / Quivi mi misi a far baratteria: / Di che rendo ragione in questo caldo'" (365). In this citation, Marinella abridges *Inferno* 22: "I' fui del regno di Navarra nato. / Mia madre a servo d'un segnor mi puose, / che m'avea generato d'un ribaldo, / distruggitor di sé e di sue cose. / Poi fui famiglia del buon re Tebaldo; / quivi mi misi a far baratteria, / di ch'io rendo ragione in questo caldo" (*Inf.* 22.48–54). She leaves out the lines about Ciampolo's mother and his conception, likely to avoid any discussion that might possibly denigrate women. The astute reader of Dante can then easily see how she is picking and choosing verses that suit her purpose. In the next line, she turns back to the Italians, or rather to Sardinia, this time to frate Gomita, of whom she writes: "E altrove dice di Gomita gran barattiere tai parole 'Denar si tolse, e lasciolli di piano / Sì come è, dice, e ne gli altri offici anche / Barattier fu no picciol, ma sovrano'" (365). Marinella has cited from the episode of frate Gomita before, in her discussion of the masculinization of *donna*. The full passage from *Inferno* reads: "Ed ei rispuose: 'Fu frate Gomita, / quel di gallura, vasel d'ogne froda, / ch'ebbe i nemici di suo donno in mano, / e fé sì lor, che ciascun se ne loda. / Danar si tolse e lasciolli di piano, / sì com' e' dice; e ne li altri office anche / barattier fu non picciol, ma sovrano" (*Inf.* 22.81–87). Frate Gomita is not actually met by Dante, but, just as Marinella indicates, is spoken of by Ciampolo. Marinella is thus again both displaying her knowledge of Dante and pulling out relatively local examples to her, geographically speaking.

Adding to the long list of Dantean defective men, Marinella makes other Dantean associations, but she indicts these men without mentioning specific citations. Can della Scala is an example of one of those who is violent to family members (385; he is also mentioned at 217). She also mentions the seditious Guelfs and Ghibellines (415–16) as well as Farinata degli Uberti (416–17).She furtherindicts the two young men of the Cancellieri family (418). These references clearly treat Dante's Florence, albeit not explicitly quoting from the poem.

I now turn to good advice that comes from a quote Marinella borrows from Dante but reinterpreted to apply to her own writing. In the sixth chapter, *Degli Otiosi, Negligenti, & Sonnacchiosi,* Marinella adds two citations of Dante: "Et chi è colui, che per il mezzo dell'otio si facci immortale? Come ben dice Salustio, e Dante: 'Che seggendo in piume. / In fama non si vien, né sotto coltre,' Certo non si può vedere la maggior infelicità di uno ingegno otioso, il quale non può sentir quel verso di Dante: 'Ratto, che'l tempo non si perda'. Ond'io spinta dalle sue parole, voglio esser breve circa questi otiosi, i quali non vogliono affaticarsi un'hora, se credessero di vivere eternamente gloriosi" (232).[26] Marinella puts two citations of Dante here: the first, from *Inferno* 24, comes from Virgil's admonishment to Dante who has sat down, tired from the climb in Malebolge, to instead continue his journey: "'Omai convien che tu così ti spoltre,' / disse'l maestro, 'ché seggendo in piuma, / in fama non si vien, né sotto coltre'" (*Inf.* 24.46–48). Both citations are commands to pick up the pace. The second citation, "'Ratto, ratto, che'l tempo non si perda / per poco amor,' gridavan li altri appresso, / 'che studio di ben far grazia rinverda.'" (*Purg.* 18.103–05), is the only Dantean citation that Marinella repeats from Passi (1599: 260), but here Marinella shows herself again to be a far more astute reader of Dante's. Passi simply cites the lines as an example against the *otiosi*. While the Dantean context sees the souls themselves crying to make haste, Marinella uses this citation for her own means and adapts the original usage to specifically focus on her own writing, wishing to be brief regarding this chapter on the slothful. She has again adapted the Dantean to the female; this is her original application, not found in Passi, and through it, Marinella's understanding of the Dantean message to run through—the writing on—sloth shines through.

In Marinella's effort to elude misogyny, she takes part in a determined effort to see the male as defective. Part of that determined effort is reading Dante as a historian—specifically one whose *terzine* categorizes defective men in a scientific or factual manner. We can thus add historicizing (or revising) Dante as another literary tool by which Marinella seeks to "controbattere la misoginia aristotelica" (Cagnolati 2013: 59). Just as Marinella selects particular passages of Aristotle's to combat while ignoring others, or while conveniently side-stepping the whole (Jordan 1990: 257), so too, in defense, does she select and reappropriate passages of Dante. Thus, along with her use of Plato (Cagnolati 2013: 59–61), Marinella's Dante becomes an asset in her toolbox, so to speak. We may thus interpret Marinella, as a reader of Dante, as one who proto-historicizes, anticipating perhaps Barolini's direction for the field of Dante studies to "Only Historicize" and to consider the material aspects of Dante's text (2022b). This is part of a larger thrust on Marinella's part, which sees historicism not only linked to feminism, but that challenges the presumption of facts or puts the truth into symbiotic functioning with historical contexts (Jordan 1990: 258). Read in this way, Marinella makes, of Aristotelian sex and gender, opinion, not science. Thus, Marinella takes, as Jordan puts it, a "historicist view of science that represents all knowledge of nature as 'opinion,' changing with time and, presumably the interests of philosophers" (260). Indeed, Paola Malpezzi Price agrees: "Marinella wants to persuade the reader that the ancient scholars' *dicta* do not constitute universal truths but rather represent opinions fashioned by the writers' personal experiences and historical situation. This idea, already suggested in Christine de Pisan's *La cité des dames* two centuries before, is implied in Marinella's attacks against Aristotle, Passi, and other detractors of the female sex" (1994: 239). Malpezzi also describes Marinella's strategy in terms of Luce Irigaray (1985), who encourages women to assume the "'mimetic' role historically imposed on them" (2008: 118–19). Thus, what can be read as Marinella's reappropriation of male literary strategies is part of a larger female genealogy that rereads male philosophers. To these stratagems, I add her historicizing use of Dante's *scienza*, or rather, her revisionist use of Dante that elevates his verse above the historians, to her arsenal of scientific opinion and philosophical argument. Thus, Dante is part of Marinella's reclaiming philosophical and "scholarly language" to aid her own writing (Malpezzi Price 2008: 119). Indeed, the use of Dante as *more* than a historian shows Marinella's refusal to elevate the *dicta* or methods of traditional philosophers and historians; she will use the vernacular poet instead. She makes his shades, his punishments, into real indictments of men. Her Dante, her interpretation of Dante, is enough.

Marinella also reappropriates the central fallacy of the male gaze which is to read the (disabled) exterior as an index of the interior. Dante complicates this binary while recognizing the deep human desire to try to read the character on the body. The body or soul whose essence or nature or signifier is deformity would be for Aristotle and Giuseppe Passi, the female body; for Dante, perhaps one example is the soul of Manfred.[27] For Marinella, it is undoubtedly the male body; she twists the stratagem for her own purposes, and while her interpretations transform sources, her bodies transform souls. This discourse on the defects of men leads us to a semiotics of the body that illuminates how gender, disability and the body are entwined with the essence of the human, of the soul. By examining Dante as a precursor to Marinella's own thought, and particularly Dante's scientific understanding of defect, I hope to have clarified how Marinella ties gender to the soul in a literary act that protects the humanity of women. Similarly, Dante consciously separates disability from the soul, perhaps, we can only speculate, with the intention to make visible the humanity of the disabled. The fundamental qualities of humanity for Dante and for Marinella as one of his Renaissance reader are qualities worth protecting, and this act of protection raises and transforms the literary stakes.

Notes

1. This idea of defect as desire is connected by Dante to love and to God. As Barolini writes in the *Undivine Comedy*, "Desire is

defective, while the cessation of desire is happiness, beatitude, in a word perfection" (1992: 26). In the poem *Donne ch'avete intelletto d'amore*, Dante uses desire as defect even in heaven without his lady: "Lo cielo, che non have altro difetto / che d'aver lei" ["Heaven, whose only imperfection is / the lack of her"] (lines 19-20). The *difetto* here is essentially the lack of having Beatrice—and as Barolini notes, it is an astonishing claim to imagine a heaven with defect. This defect is not framed as inherently negative. For geographic "defects," see also Barolini (2022: 22-44).

2. Dante uses the term "difetto" eight times in the *Commedia*; three times in *Inferno;* three times in *Purgatorio;* twice in *Paradiso*. In *Inferno*, the first use is the defect of having lived before Christ; the second is the defective judgment of the devils who are tricked by Ciampolo; the third is Dante's own defective judgment, as admonished by Virgil. In *Purgatorio*, the first is defect as sin which is not amended by prayer; the second sees all humans as defective; the third sees Forese Donati's lack of *carne*, of flesh, as a defect. Finally, in *Paradiso*, defective reasoning is a trait universal to humans; defect then returns to Dante, as admonished by Beatrice. The uses of *difetto* often do not refer directly to the disabled body in the *Commedia*. However, the occurrences of *difetto* in the *Commedia* show a semantic area in which physical defects are also a part. This approach emerges further in the *Convivio*.
3. See also Dailey (2013: 2): "In the Christian tradition, the body is not conceived of as a simple organic unity, but rather as a twofold entity partaking of two anthropological registers—the inner and the outer persons—that promises to find its true materiality in a time to come." The relationship between these two persons is especially considered here when the "outer person" is disabled (or female).
4. This discussion of the (deformed) body of course owes much to the Aristotelian relation between form and matter, embryology, and the formation of the body.
5. What Dante signifies by the nebulous "loro simili" is a point of inquiry in my ongoing work. "Muteness" was historically associated with deafness, particularly by medieval writers following Aristotle (*Historia animalium* 536b) and Pliny (*Historia naturalis* 10.88).

6. All citations of Marinella come from Lucrezia Marinelli, Francisco Guerra, and Giovanni Battista Ciotti, *La nobilta, et l'eccellenza delle donne, co' diffetti, et mancamenti de gli huomini*, Ricorretto, accresciutto in questa seconda impressione (In Venetia: appresso Gio. Battista Ciotti sanese, all'insegna dell'Aurora, 1601), herafter cited only with page numbering.
7. Marinella states her stance on female vs. male beauty clearly: "le donne adunque sono più belle de gli huomini, che per lo più sono rozzi, e mal composti" (25). Marinella's approach to beauty will make it a virtue, one that leads to the divine. Further, she shifts from Plato's male register of beauty to a female one. Marinella makes beauty a good, which Passi had as an evil: "Giuseppe Passi, il quale scrivendo alla cieca, ardisce di affermare che la beltà sia cagione d'infiniti mali" (21).
8. Baradel and Zancan (1983: 104) refer to this work as a "semplice repertorio di *exempla*."
9. As Gragnolati argues, the popular debate between plurality and unity of form reaches a crux in Dantean embryology (as explicated in *Purgatorio* 25). Dante conceives of the embryo as a combination of Aquinas's discontinuous process and Bonaventure's continual evolution from one soul to the other. Dante follows the principles of Thomism up to the formation of the sensitive soul. It is still a singular soul, but created differently.
10. Dante's lack of connection between vice and disability owes more to the pluralists than to Aquinas. How much Dante followed Aquinas in the connection of physical impairments to sin is up for debate; Bruno Nardi cites Dante's dependence on Albert regarding the origin of *umana bontade* or nobility. Nardi also claims Dante follows Albert, too, in the conception of the soul: "partim ab intrinseco et partim ab extrinseco," i.e. partly generated by nature and partly by God (1931). But Gragnolati is more specific: "Dante's original and ambivalent theory begins by following the empirical tenets of plurality (which maintain the concreteness of the body and clearly distinguish it from the soul) but ends up being close to the idea of unicity of form (which considers the rational soul to be the full carrier of the self, including the body)" (2005: xv). Dante's conception of the soul may appear similar to the unicity model of Aquinas, but because he follows the tenets of plurality—of which Bonaventure was a proponent—that elevate the

importance of the body, I am unconvinced that, for Dante, sin and physical disability were necessarily tied. For Dante, we must ask whether the deformed body was direct evidence of a deformed soul. Trovato (1990) sees the encounter with Statius as Dante's attempts to correct opinions regarding the nature of the intellectual soul. Dante was countering the thought coming out of Bologna—which was, at that time, predominantly Averroistic. Dante follows Albert, who defined the intellectual soul as the most noble form in nature: the purer the recipient, the more intellect shines within them. On the same question, see also Lindheim (1990), and Boyde (1993).

11. Aristotle's work arguably had the largest effect on Dante's own. For Aristotle, the connection between body and soul was well worth investigating. Not every body was ensouled, and matter could be independent of the soul. It is not that there is a body and then a defective soul which makes the body disabled. There is no soul independent of the body, thus it is unclear how Aristotle would explain the disabled body, since we cannot think of soul without body. The tradition of a Christian interpretation of Aristotle (as in Aquinas) did not conflict with the doctrine of the immortality of the soul—the separable mind in *De anima* has a role in this interpretation.
12. The bibliography on the subject is immense. Among others, see Callus (1960: 581–82): "The soul is in the strict sense the substantial form of the body whereby it is what it is: the flesh, the veins, the nerves, the muscles and other bodily members are not human except in virtue of the soul."
13. This argument complicates Deslauriers' claim that "In referring to the division of the soul, Marinella does not distinguish between Platonist and Aristotelian accounts, assert their harmony, or point out divergences" (721).
14. The whole paragraph reads: "Nel sostenere, infatti, la presenza nella mente di Dio dell'idea della donna e, simultaneamente, l'esistenza di un'anima femminile, differente da quella maschile, Lucrezia Marinella delegittima il modello maschile come unico possibile per l'essere umano e riconosce la donna in quanto tale come piena manifestazione dell'umano."

15. As stated in Baradel and Zancan (1983), "Lucrezia Marinella ci informa sulle letture preferite, sulla diffusione di certi testi e naturalmente sulla loro maneggiabilità da parte del nuovo pubblico femminile" (105).
16. Commenting *Inf.* 4.139-40 in his *Esposizioni* (1994), Boccaccio in fact writes "delle femine le quali quel natural difetto patissono, del quale esse ogni mese sono, almeno una volta, impedite" (205).
17. Marinella's title page reads that in this work "non solo si distrugge l'opinione del Boccaccio, d'amendue i Tassi, dello Sperone, di Monsignor di Namur, e del Passi; ma d'Aristotele il grande." Kolsky (2001) notes that Passi is last of the modern writers possibly due to Marinella's "realization that there was a clearly delineated Italian genealogy of misogynous thought" (978).
18. On Aristotelian biology and gender, especially the statements presented in *De generatione animalium* 2.1, 732a 2-9 and *Politica* 7.16, 1335a 11-17, see Connell (2016); Merriam (2010); Rose (2003). Aristotle discusses deformity in a number of ways. In the *De Generatione Animalium*, the second recorded work on embryology, it is used both to refer to baldness and to women. Connell writes that in the *Generation of Animals*, "the female is both functional and a failure" (2016: 280). While Aristotle describes the female as a deformed male, he takes this approach based on his own sense of biology, because females cannot produce male sperm. However, females are necessary for the act of reproduction. Aristotle's system of biology certainly shaped his understanding of gender. In particular, he made these statements on women as deformed based on his understanding of species essentialism. In practice, for women in the ancient world, sterility was a more serious matter than disability. For evidence on the causes of deformity, Aristotle cautions that women who are too young may have children who are not properly developed. Similarly, he cautions against men who are too old. His concern is also for young women who suffer more in childbirth. He further advises that expecting mothers should exercise, but they should keep their minds idle. These passages put forth some ideas on why bodies might be weak or incomplete—when the parents are too young or too old; when women exercise their minds. Aristotle is giving the reader some information regarding what causes deformity, or weakness.

19. While Marinella agrees with Aristotle in using temperature as an indicator of character, she asserts instead, following Galen, that women's temperate nature makes them more stable. See also Deslauriers (2019: 725); Benson (1992: 29); D'Alessandro Behr (2018: 90–91).
20. See Deslauriers (2019: 726): "In appealing here to Ficino, Marinella is treating temperature as a physical instrument of the immaterial soul, and an instrument that varies in its adequacy—either extreme of temperature affects its performance negatively. So moderate heat, of the degree found in the cooler bodies of women, is the best instrument of the soul, producing positive effects that are both intellectual and moral."
21. See Allen and Salvatore (2009: 17): "Appealing to the authority of Aristotle who claimed, in the eighth book of the *Metaphysics*, that proper nouns reveal something of the nature and essence of things, Marinelli identifies five nouns that are particularly suited to comprehend the origin of 'woman'." Marinella's discussion of "donna" is indebted to Agrippa, and she is also responding to Passi's own evocation of the topoi of *donna* as *danno,* to which Moderata Fonte also responds; Fonte indicated *donna* as a *dono,* instead (Panizza 1999; Baradel and Zancan 1983: 128; Passi 1599: 974).
22. The main difference between Aristotle and Plato is that the soul is not separable from the body in Aristotle. In Plato, the soul is immortal; it has to purify itself from the body and will live on and have another body. In the *De anima*, Aristotle puts forth a relation between form and matter: there is no soul without a body: "the soul neither does anything nor has anything done to it without the body" (*De anima* 1.1, 403a). He repeats this: "the attributes of the soul are inseparable from the natural material of living things" (1.1, 403b 10). For Aristotle, "the soul is that which produced motion" (1.2, 403b 20). This concept of Aristotelian motion is based, to be very concise, on the idea of potentiality and actuality. The powers of the soul include "nutrition, perception, motion with respect to place, and thinking things through" (2.2, 414a 30). The body is the medium through which the soul actualizes its potentialities. And the soul, defined by three qualities: "motion, sense perception, and bodilessness" (1.2, 405b 10), is that which "holds the body together; at any rate, when the soul has gone

out of it, the body gives off vapors and rots away" (1.2, 411b). According to Aristotle, there are different forms or functions of the soul: the vegetative (plants, animals and humans), the sensitive (animals and humans) and the rational (humans alone). His interest is biological and epistemological. For Aristotle, the functions of the body are inextricably connected to those of the soul (imagination, memory, and sense-perception). The soul is the actuality of the body, the form, and so the two are inextricable. In his ethical texts, Aristotle goes further to distinguish between two parts of the soul, one rational and the other irrational; then good and virtuous behavior depends on the good, rational part of the soul mastering the irrational passions.

23. The *Dantista* might hope for an invocation of Beatrice *loquax*, but Marinella sticks with only male examples from Dante's oeuvre in this second half of her work. See Barolini (2006: 360-78).
24. The Dantean shades only look like bodies. Disability and the corrupted body are such complex issues within the time frame and within Dante's afterlife; it becomes nearly impossible to separate the realities of living from the theorized spiritual eventuality of the body being restored with the soul. On his journey, Dante is intrigued by the abilities, composition, and fate of the souls he encounters. In *Canto* 6, Virgil explains that each soul will regain "sua carne e sua figura" (*Inf.* 6.98) at the Last Judgment. Yet this restoration of the body will only increase the souls' suffering in hell. One of the most vivid passages that addresses the fate of the souls' bodies is found in *Inferno* 7, in which it is said the souls will be resurrected: "Col pugno chiuso, e questi coi crin mozzi" (*Inf.* 7.57). In fact, the only souls to whom their bodies will not be restored are the suicides. The suicides will never regain their humanity—their bodies forever separate, dangling from the tree of their soul. Faced with these and other aberrations of the flesh, Dante moves heavily through hell; his body is real and presumably relatively able. See Gragnolati (2003; 2005).
25. Marinella writes: "Però dice il Petrarca; / 'E vidi: Ciro più di sangue avaro, / che Crasso d'oro e l'uno, e l'altro n'hebbe / Tanto a la fin, ch'a ciascun parve amaro'" (188).
26. Sallust also writes of *otium*. See Bruce (2021).
27. In Dante's treatment of disability in the afterlife, one might consider the presence of Manfred's scars in *Purgatorio* 3. A scar

does not necessarily constitute a disability, but I take this particular visual sign as an initial example of what unique bodily features persist in the afterlife. The episode of Manfred has caught many commentators' attention. See Freccero (1986); Balfour (1993).

References

Alighieri, Dante. 2011. *Convivio*, ed. Gianfranco Fioravanti and Claudio Giunta. In *Opere*, vol. 2: *Convivio, Monarchia, Epistole, Egloge*, ed. Marco Santagata, 2 vols. Milan: Mondadori.

Alighieri, Dante. 2016. *Commedia*, ed. Anna M. Chiavacci Leonardi. Milan: Mondadori.

Allen, Prudence, and Filippo Salvatore. 2009. Lucrezia Marinelli and Woman's Identity in Late Italian Renaissance. *Renaissance and Reformation* 28 (4): 5–39.

Aristotle. 1984. *The Complete Works of Aristotle: The Revised Oxford Translation*, ed. Jonathan Barnes, 2 vols. Princeton, NJ: Princeton University Press.

Aristotle. 2000. *Politics*. Transl. B. Jowett. Mineola, NY: Dover Publications.

Aristotle. 2014a. *Generation of Animals*. Transl. A.L. Peck. Cambridge, MA: Harvard University Press.

Aristotle. 2014b. *History of Animals*. Transl. D.M. Balme, and A.L. Peck, 3 vols. Cambridge, MA: Harvard University Press.

Balfour, Mark. 1993. 'Orribil Furon Li Peccati Miei': Manfred's Wounds in *Purgatorio* III. *Italian Studies* 48 (1): 4–17.

Baradel, Virginia, and Marina Zancan, eds. 1983. *Nel cerchio della luna: figure di donna in alcuni testi del XVI secolo*. Venice: Marsilio.

Barolini, Teodolinda. 1992. *The Undivine Comedy: Detheologizing Dante*. Princeton, NJ: Princeton University Press.

Barolini, Teodolinda. 2006. *Dante and the Origins of Italian Literary Culture*. New York: Fordham University Press.

Barolini, Teodolinda, ed. 2014. *Dante's Lyric Poetry: Poems of Youth and the Vita Nuova (1283–1292)*. Buffalo: University of Toronto Press.

Barolini, Teodolinda. 2019. Dante's Limbo and the Cultural Other, or: Injustice on the Banks of the Indus. In *Dante Worlds: Echoes, Places, Questions*, ed. Peter Carravetta, 21–34. Rome: L'Erma di Bretschneider.

Barolini, Teodolinda. 2022. "'Only Historicize': History, Material Culture (Food, Clothers, Books), and the Future of Dante Studies." In *Dante's Multitudes: History, Philosophy, Method*, 3–21. Notre Dame: University of Notre Dame Press.

Barolini, Teodolinda. 2022. Dante's Sympathy for the Other, or the Non-Stereotyping Imagination: Sexual and Racialized Others in the *Commedia*. In *Dante's Multitudes: History, Philosophy, Method*, 22–44. Notre Dame: University of Notre Dame Press.

Benson, Pamela. 1992. *The Invention of the Renaissance Woman: Female Independence in the Literature and Thought of Italy and England*. University Park: Pennsylvania State University Press.

Boccaccio, Giovanni. 1994. *Esposizioni sopra la Comedia di Dante*, ed. Giorgio Padoan. Milan: Mondadori.

Boyde, Patrick. 1993. *Perception and Passion in Dante's Comedy*. New York: Cambridge University Press.

Bruce, Keegan. 2021. *Shifting Discourses of Roman Otium in Cicero, Catullus, and Sallust*. MA thesis. The University of Western Ontario. https://ir.lib.uwo.ca/etd/7718. Accessed 29 June 2023.

Cagnolati, Antonella. 2013. Un duello in punta di penna: Strategie antimisogine nella *Nobiltà*. In *A Portrait of A Renaissance Feminist: Lucrezia Marinella's Life and Works*, ed. Antonella Cagnolati, 41–66. Rome: Aracne.

Callus, Daniel A. 1960. The Problem of the Plurality of Forms in the Thirteenth Century: The Thomist Innovation. In *L'homme et son destin, d'après les penseurs du moyen âge. Actes du premier Congrès international de philosophie médiévale*, ed. Marie-Dominique Chenu, 577–585. Louvain and Paris: Nauwelaerts.

Cassell, Anthony K., ed. 2004. *The Monarchia Controversy: An Historical Study with Accompanying Translations of Dante Alighieri's Monarchia, Guido Vernani's Refutation of the Monarchia Composed by Dante and Pope John XXII's Bull Si Fratrum*. Washington, DC: The Catholic University of America Press.

Cerbo, Anna. 2001. *Poesia e scienza del corpo nella* Divina commedia*: Dicer del sangue e de le piaghe*. Naples: Libreria Dante & Descartes.

Chemello, Adriana. 2000. The Rhetoric of Eulogy in Lucrezia Marinella's *La nobiltà et l'eccellenza delle donne*. In *Women in Italian Renaissance Culture and Society*, ed. Letizia Panizza, 463–474. Oxford: Legenda.

Connell, Sophia M. 2016. *Aristotle on Female Animals: A Study of the Generation of Animals*. Cambridge: Cambridge University Press.

Cox, Virginia. 2008. *Women's Writing in Italy, 1400–1650*. Baltimore: Johns Hopkins University Press.

Cox, Virginia. 2011. *The Prodigious Muse: Women's Writing in Counter-Reformation Italy*. Baltimore: Johns Hopkins University Press.

Dailey, Patricia. 2013. *Promised Bodies: Time, Language, and Corporeality in Medieval Women's Mystical Texts*. New York: Columbia University Press.

D'Alessandro Behr, Francesca. 2018. *Arms and the Woman: Classical Tradition and Women Writers in the Venetian Renaissance*. Columbus: The Ohio State University Press.

De Tommaso, Emilio Maria. 2020. Della natura et essenza del donnesco sesso: Ontologia della differenza di genere in Lucrezia Marinella. *Bruniana e Campanelliana* 24 (1): 63–75.

Deslauriers, Marguerite. 2017. Marinella and Her Interlocutors: Hot Blood, Hot Words, Hot Deeds. *Philosophical Studies: An International Journal for Philosophy in the Analytic Tradition* 174 (10): 2525–2537.

Deslauriers, Marguerite. 2019. Patriarchal Power as Unjust: Tyranny in Seventeenth-Century Venice. *British Journal for the History of Philosophy* 27 (4): 718–737.

Fonte, Moderata. 1988. *Il merito delle donne: ove chiaramente si scuopre quanto siano elle degne e più perfette de gli uomini*, ed. Adriana Chemello. Milan: Eidos.

Freccero, John. 1986. Manfred's Wounds and the Poetics of the *Purgatorio*. In *Dante: The Poetics of Conversion*, ed. Rachel Jacoff, 195–208. Cambridge, MA: Harvard University Press.

Giallongo, Angela. 1995. *L'avventura dello sguardo: educazione e comunicazione visiva nel Medioevo*. Bari: Edizioni Dedalo.

Gilson, Etienne. 1967. Dante's Notion of a Shade: *Purgatorio* XXV. *Mediaeval Studies* 29 (1): 124–142.

Godden, Richard H., and Asa Simon Mittman, eds. 2019. *Monstrosity, Disability, and the Posthuman in the Medieval and Early Modern World*. Cham: Springer Nature.

Gragnolati, Manuele. 2003. From Plurality to (Near) Unicity of Forms: Embryology in *Purgatorio* 25. In *Dante for the New Millennium*, eds. Teodolinda Barolini, and Wayne Storey. New York: Fordham University Press.

Gragnolati, Manuele. 2005. *Experiencing the Afterlife: Soul and Body in Dante and Medieval Culture*. Notre Dame: University of Notre Dame Press.

Granacki, Alyssa M. 2021. 'Molti e Molte:' Women Readers and Lady Philosophy in Dante's *Convivio*. *Bibliotheca Dantesca* 4 (1): 50–66.

Irigaray, Luce. 1985. *This Sex Which Is Not One*. Transl. C. Porter, and C. Burke. Ithaca, NY: Cornell University Press.

Jordan, Constance. 1990. *Renaissance Feminism: Literary Texts and Political Models*. Ithaca, NY: Cornell University Press.

Jordan, Constance. 1996. Renaissance Women Defending Women: Arguments against Patriarchy. In *Italian Women Writers from the Renaissance to the Present: Revising the Canon*, ed. Maria Ornella Marotti, 55–67. University Park: Pennsylvania State University Press.

Kolsky, Stephen. 2001. Moderata Fonte, Lucrezia Marinella, Giuseppe Passi: An Early Seventeenth-Century Feminist Controversy. *The Modern Language Review* 96 (4): 973–989.

Lindheim, Nancy. 1990. Body, Soul, and Immortality: Some Readings in Dante's *Commedia*. *MLN* 105 (1): 1–32.

Malpezzi Price, Paola. 1994. Lucrezia Marinella (1571–1653). In *Italian Women Writers: A Bio-Bibliographical Sourcebook*, ed. Rinaldina Russell, 234–242. Westport: Greenwood Press.

Malpezzi Price, Paola. 2008. *Lucrezia Marinella and the Querelle Des Femmes in Seventeenth-Century Italy*. Madison: Fairleigh Dickinson University Press.

Marchesi, Simone. 1999. 'Epicuri de Grege Porcus': Ciacco, Epicurus and Isidore of Seville. *Dante Studies* 117: 117–131.

Marinella, Lucrezia. 1999. *The Nobility and Excellence of Women, and the Defects and Vices of Men*, ed. Anne Dunhill. Chicago: University of Chicago Press.

Marinelli, Lucrezia, Francisco Guerra, and Giovanni Battista Ciotti. 1601. *La nobiltà, et l'eccellenza delle donne, co' diffetti, et mancamenti de gli huomini. Ricorretto, accresciutto in questa seconda impressione*. Venice: appresso Gio. Battista Ciotti sanese, all'insegna dell'Aurora.

Mazzeo, Joseph Anthony. 1957. The Analogy of Creation in Dante. *Speculum* 32 (4): 706–721.

Merriam, Garret. 2010. Rehabilitating Aristotle: A Virtue Ethics Approach to Disability and Human Flourishing. In *Philosophical Reflections on Disability*, eds. D. Christopher Ralston, and Justin Hubert Ho, 133–151. Dordrecht: Springer Netherlands.

Nardi, Bruno. 1931. L'origine dell'anima umana secondo Dante. *Giornale Critico Della Filosofia Italiana* 12: 433–456.

Panizza, Letizia. 1999. Introduction. In Lucrezia Marinella, *The Nobility and Excellence of Women, and the Defects and Vices of Men*, ed. Anne Dunhill. Chicago: University of Chicago Press.

Passi, Giuseppe. 1599. *I donneschi diffetti, nuovamente formati e posti in luce*. Venice: Appresso Iacobo Antonio Somascho.

Plastina, Sandra. 2017. Nella milizia riuscir più d'una: Amazzoni e donne in armi da Boccaccio a Lucrezia Marinella. *Bruniana e Campanelliana* 23 (2): 493–503.

Pliny the Elder. 1940. *Natural History, vol. 3: Books 8–11*. Transl. H. Rackham. Cambridge, MA: Cambridge University Press.

Ray, Meredith K. 2015. *Daughters of Alchemy: Women and Scientific Culture in Early Modern Italy*. Cambridge, MA: Harvard University Press.

Rose, Martha L. 2003. *The Staff of Oedipus: Transforming Disability in Ancient Greece*. Ann Arbor: University of Michigan Press.

Ross, Sarah Gwyneth. 2009. *The Birth of Feminism: Woman as Intellect in Renaissance Italy and England*. Cambridge, MA: Harvard University Press.

Schnieders, Laura. 2014. La polemica dei sessi: Lucrezia Marinella e Giuseppe Passi. In *Conflitti culturali a Venezia dalla prima età moderna ad oggi*, eds. Rotraud von Kulessa, Daria Perocco, and Sabine Meine, 191–206. Florence: Cesati.

Shapiro, Lisa. 2013. The Outward and Inward Beauty of Early Modern Women. *Revue Philosophique De La France Et De L'étranger* 203 (3): 327–346.

Sinclair, Amy. 2020. *Insinuatio* in Lucrezia Marinella's *Essortationi Alle Donne* (1645): Exhorting Marital Harmony and Insinuating Feminist Critique. *Renaissance Studies* 34 (3): 430–446.

Tarabotti, Arcangela. 2007. *La semplicità ingannata: edizione critica e commentata*, ed. Simona Bortot. Padua: Il poligrafo.

Trovato, Mario. 1990. Dante's Stand against 'l'errore de l'umana Bontade': *Bonum*, Nobility and Rational Soul in the Fourth Treatise of the *Convivio. Dante Studies* 108 (1990): 79–96.

Ureni, Paola. 2010. Human Generation, Memory and Poetic Creation: From the *Purgatorio* to the *Paradiso. Quaderni D'italianistica* 31 (2): 9–33.

Weiss, Sonja. 2022. Ficino's Doctrine of Love and Beauty and Its Plotinian Background. *Filozofia* 77 (1): 20–35.

Willer, Annika. 2013. Silent Deletions: The Two Different Editions of Lucrezia Marinella's *La nobiltà et l'eccellenza delle donne. Bruniana e Campanelliana* 19 (1): 207–219.

CHAPTER 7

Speaking Like a Philosopher: Dante and Renaissance Treatises on Love

Eva Del Soldato

John Monfasani wrote an article a few years ago, which he opened with a quotation from a Sherlock Holmes adventure, the *Silver Blaze*, about a dog that did not bark (1993: 247). It was that event that did not happen, indeed, to be relevant and, ultimately, to allow Holmes to solve the case. In the same way, Monfasani capitalized on the absence of Ockhamists in Pre-Reformation Italy to investigate the degree of philosophical liberty in that period—especially on crucial issues like the immortality of the soul. This chapter also focuses on the absence, or at least on a circumscribed and qualified presence, of Dante in the Renaissance love treatises. Discussing why Dante was used and invoked in a limited way in a genre that was so popular in the Renaissance period, this chapter will offer insights about the authoritative function(s) attributed to Alighieri in

E. Del Soldato (✉)
Department of Francophone, Italian, and Germanic Studies, University of Pennsylvania, Philadelphia, PA, USA
e-mail: evadel@sas.upenn.edu

M. Pace (ed.), *Dante and the Sciences of the Human*, Palgrave Studies in Medieval and Early Modern Medicine,
https://doi.org/10.1007/978-3-031-69253-6_7

sixteenth-century Italy and especially in Florence, building upon previous contributions, including Simon Gilson's recent monograph (2018).

Renaissance treatises on love, including those dealing more specifically with lovesickness, were part of a genre with noticeable continuities, to the extent—at times—to appear even dull and repetitive, with very few shifts from one author to another. In terms of sources and authorities, Platonic and neo-Platonic templates were regularly applied with an impressive lack of originality, and the use of Aristotle was often limited to references to the chapters on friendship in the *Nicomachean Ethics*.[1] This repetitiveness was—among other things—particularly visible in the display of poetic authorities quoted in love treatises, as made evident also by those works that were expressly conceived as commentaries on love poems.

Among the quoted poetic authorities, the lion's share belonged certainly to Petrarch: in many instances, he was barely called by his name but referred to as "the poet" and competed with Plato in terms of mentions (Ebbersmeyer 2002: 45; Favaro 2012: 8). Petrarch was evoked when authors discussed how one can re-orient human love toward a higher, divine object, or how love can persist after the death of the beloved. Yet, the self-analysis of the *Canzoniere*, which describes the poet's feelings for Laura in painstaking detail, and also the *Trionfi*, which also made available a complete symptomatology of love, was particularly useful when there was a need to discuss its physical effects. Commentaries and lectures on Petrarch's poems on love became particularly common after the 1540s, with the rise of the age of *Accademie*.

A tradition of its own—spanned over three centuries—is represented by the commentaries on Guido Cavalcanti's *Donna me prega*. One can think of the works by Iacopo Mini, Paolo Del Rosso, Francesco Verino Secondo, and Girolamo Fracchetta, authors who continued and revisited the practice started by the physician Dino del Garbo and the Ps.-Aegidius in the fourteenth century, reading the text and its treatment of love according to medical, philosophical and literary perspectives.[2] Guido and his *canzone* had also been central to Marsilio Ficino's commentary on Plato's *Symposium*, the *Libro de amore* (1474, Latin version 1469), so provoking Pico della Mirandola's reaction, in the *Commento sopra una canzona de amore di Girolamo Benivieni* (1486).[3]

Having mentioned Petrarch and Cavalcanti, one could wonder about the place of Dante in the love treatises's tradition. There is undoubtedly an intertextual presence of Dante's *Commedia* in the love treatise that represents—in many respects—the blueprint for many subsequent ones, that is precisely Ficino's *Libro*.[4] In the dedicatory letter to the vernacular version of the text, Ficino suggested the parallelism of his own text with Dante's *Commedia* through a web of easily identifiable references (the "dark forest," the "straightforward pathway": Gilson 2009: 145–46), implying that both works aimed at teaching to love "rettamente." The Platonic shadow Ficino cast on Dante went beyond that: it can be perceived in the decision of sixteenth-century editors of Dante's *Convivio* to print the work under the title of *L'amoroso convivio*, establishing, in this case, a parallel with Plato's *Symposium* itself, as shown by Cesare Vasoli (Alighieri 1988: 1. lxxxi–lxxxii). Yet, in the pantheon of authorities used in love treatises, Dante nonetheless seems to occupy a secondary role, as a sort of background actor.

In fact, already in Ficino's *Libro,* Dante's presence was merely alluded to, unlike Cavalcanti's. Moreover, Ficino's broader interest in Dante was closely intertwined with his linguistic program, focused on the elevation of the Florentine language, which had a primarily local appeal. One could be tempted to speculate that the sixteenth-century canonization of Petrarch, under the auspices of Bembo, was therefore decisive to sideline Dante in this context: after all, love treatises were typically composed in the vernacular and an excellent vehicle for its advancement. This explanation is nonetheless insufficient because many of Bembo's opponents in the *questione della lingua* equally granted Petrarch the role of central authority when writing and speaking of love. One could consider Mario Equicola (1526), who rejected both Bembo and Florentinism, but supported a *comune italica lingua* and celebrated the newly re-discovered *De vulgari eloquentia* as an ally (1526: 155r). In the printed and final version of his *Libro de natura de amore* (first edition 1525, corrected version 1526), Petrarch is quoted more often than Dante, and always in more substantial ways. At one point, when explaining that no one suffered more than Petrarch for the death of the beloved, Equicola directly compares him with other poets, including Dante, and claims that Petrarch showed more grief for Laura than Dante ever showed for the passing of Beatrice: in his "*vita novella*" Dante did not "narrate the anguish" for Beatrice's death as much as Petrarch did for Laura's in his verses (1526: 9v). In a few words, Petrarch offered a spectacle of love to which Dante's treatment could not compare.

There is some truth in Equicola's words, albeit rhetorical in their essence. If scholars like Natascia Tonelli have insisted on Dante's profound expertise in the pathologies linked to amorous passion (2015,

2021), expertise especially displayed in the *Rime* and the *Vita nuova*, they also pointed out that, in that latter work, the poet is eager to point out that his feelings for Beatrice are not a destructive passion which consumes the body and the mind of the lover. In *Vita nuova*, Beatrice is "so virtuous that she could not bear for Dante's love for her not to be counseled by reason" (*VN* 2).[5] This perspective does not prevent Dante from offering details about the physical symptoms of love that affected him, but his condition is posited on a different ground of rational control that does not convert itself into the pathological *amor hereos*. If one considers the nature of Dante's love for Beatrice, and its characterization in *Vita nuova* and in the *Commedia* (in which the connection between actual love and rationality is often and variously reaffirmed),[6] the words of Equicola seem to find some textual foothold. In works like *Vita nuova* and the *Commedia*, Dante's reason is not overwhelmed by desire, while he looks at an object of love that is ultimately beyond Beatrice herself. It seems, therefore, that Dante could not be perceived as a love poet, and thus an authority on love (and lovesickness), in the same way, as Petrarch or Cavalcanti.

Still, it must be noted that the identification of Dante as a love poet was not uncommon and, at times, carried unflattering undertones. If Petrarch had characterized Dante as one among many love poets—in *Rvf.* 287 and *Tr. Cup.* 4.28–37—in order to downgrade his predecessor, Machiavelli famously labeled Dante as much as Petrarch as love poets, to be read "for recreational pleasure" in opposition to the great classical authors from the past.[7] On the other hand, those who sympathized for Dante, starting with Boccaccio, did not perceive him as a one-trick pony in terms of subjects treated in his works and certainly did not want to reduce him to be a mere love poet. In the sixteenth century, with the *questione della lingua* that pitted Dante against Petrarch, the issue became a recurring one. The *Difesa della lingua fiorentina e di Dante* (1556)—primarily by Carlo Lenzoni, but completed by two other members of the Accademia Fiorentina, Pierfrancesco Giambullari, and Cosimo Bartoli—even featured a two-column comparison between Petrarch and Dante, where it is stated that the former "discusses few other issues, aside from love" while the latter "discuss infinite subjects."[8] In making these claims, the *Difesa* offers a decisive hint to understand the Petrarchean supremacy as an authority in love treatises: the narrow focus of Petrarch's expertise in his poetry, so to say, would explain why he could be perceived as a more obvious and suitable reference when speaking of love if opposed to Dante's encyclopedism. In his 1554 *Sette dubbi sull'amore*, a lecture for the Accademia Fiorentina, Benedetto Varchi (1503–1565) had also accepted to offer his solution to the question of whom, among Petrarch and Dante, had *better* discussed of love. In Varchi's view, it depended on the meaning assigned to the word "better": if "better" meant who discussed more extensively of love, and more gracefully, Petrarch was undoubtedly to be preferred. But if "better" meant more eruditely or more truly—as a philosopher would do—Dante had no rivals.[9]

The perception of Dante as a philosopher more than a poet was common in the sixteenth century and, as shown also recently by Gilson (2018), not always with the celebrative tones that we find in the *Difesa di Dante* and Varchi.[10] To mention a single example, in Bernardino Tomitano's *Ragionamenti della lingua Toscana* (1546), a philo-Bembian text that urged the composition of the *Difesa di Dante*, Dante is accused of having often forgotten being a poet, acting instead as a philosopher and a theologian.[11] This brings us to the Dante that does appear, albeit occasionally, in sixteenth-century treatments of love.

Sixteenth-century love treatises include quotations from Dante's *Rime* (Favaro 2021: 21, 62), but they still draw more often from the *Commedia*. And yet, one would be possibly surprised by which sections of the poem were privileged. To today's readers, *Inferno* 5 would seem the most obvious section for fishing material on love. Francesca's speech, in particular, would appear to offer solid evidence for discussions on the phenomenology of love and its destructive effects. However, a review of the presence of *Inferno* 5 in several sixteenth-century love treatises would be in many ways disappointing, being usually limited to *Inf.* 5.103 ("Amor, ch'a nullo amato amar perdona"), staged in a varied set of uses.[12] In his *Dialogo di amore*, Sperone Speroni cursorily mentioned *Inf.* 5.103 in association with *Purg.* 2.76–78 (the encounter with Casella) to praise Dante, who distinguished the reciprocity of love in men and women (1596: 22). Lodovico Dolce quoted *Inf*. 5.103—somehow surprisingly—to instruct wives that they must love their husbands to be loved back (Dolce 1560: 46v). The Brescian physician Bartolomeo Arnigio put *Inf*. 5.103 in the mouth of an unhappy lover, eager to justify his sufferings, in the *Dialogo della medicina d'amore* (Arnigio 1566: 10r–v). Flaminio Nobili, used instead *Inf.* 5.103 in opposition to Ariosto's "Ingiustissimo amor" (*Orlando Furioso* 2.1), claiming that Dante was generally right in affirming the reciprocity of love (Nobili 1590: 35v–36r).[13]

Statistically speaking, on the other hand, the predominance of quotations from Dante in love treatises comes from the second cantica, *Purgatorio*. But even here, Dante in love, Dante recognizing the "segni dell'antica fiamma," when he meets Beatrice again in *Purgatorio* 30, captured the attention of love writers much less than the central sections of the cantica, namely *Purgatorio* 17 and 18.[14]

Dante offers in *Purgatorio* 17 and 18 long definitory sections on love, which serve—among other things—as a crucial justification of the structure of *Purgatorio* itself. It is precisely in *Purgatorio* 18 that Dante states that, albeit we are naturally inclined to it, love is not an irresistible force that necessarily destroys human reason, but instead something that free will can regulate. As a matter of fact, these doctrinal sections from *Purgatorio* are those that appear more consistently in sixteenth-century treatments of love: we can already observe in a preliminary form that Dante the philosopher, Dante and his rational love—rational not only because Dante discusses it analytically but also because he does not oppose it to reason—were deemed more interesting than the effects that love would have produced on him.

Giancarlo Alfano, for example, noted that *Purgatorio* 17 and 18 are consistently mentioned in sixteenth-century commentaries on Guido's *Donna me prega*, namely in Mini, Del Rosso, and Vieri (2010). Even if some of these authors, and especially Del Rosso, invoked Dante in their commentaries often to discuss linguistic issues, their references to *Purgatorio* 17 and 18 served the purpose of emphasizing—in Alfano's view—the connection between poetry and philosophy in both Guido and Dante and therefore Florence's cultural primacy. But *Purgatorio* 17 and 18 figured preeminently already in the above-mentioned Mario Equicola. Indeed, Equicola built the section on Dante in his *Libro de amore* mostly as a collage of quotations from *Purgatorio* 17 and 18, concluded by the identification *in via Dantis* of divine love as the supreme one (1526: 7r–8r).[15] Flaminio Nobili used *Purg*. 18.19–33 to show the association between love and desire, and in a different section he paraphrased the most doctrinal verses of *Purgatorio* 18—which culminate in lines 70–72—to prove that the intellect is ultimately in control of the senses, and therefore true love is a product of our free election (1590: 16r–v, 33r–v). These examples and their contexts fittingly demonstrate that according to sixteenth-century writers Dante's authority on love was not to be sought for minute and concrete physiological descriptions of feelings and passions, but rather in abstract theoretical analyses.

But more can be added. There are, in fact, even instances in which *Purgatorio* 17 and 18 represented the exclusive focus, or at least the starting point, of dissertations on love. These treatments were conceived as public lectures at the Accademia Fiorentina, the institution that we had already seen eager to celebrate Dante not only as a poet but also as a philosopher (Plaisance 2004; Moyer 2020). As in the case of other lessons held by the *Accademici* on different subjects, these lectures on love had a formalized structure made of questions and solutions, in a scholastic style, that took as a starting point a few verses to comment on.

Perfect examples of the organization of these academic lectures are three lessons by Francesco de' Vieri or Verino Primo (1474–1541), who had the distinguished honor of reading them at the inauguration of the activities of the Accademia Fiorentina in February 1541. These lectures—published posthumously in 1547—are the only text that survived among those penned by the Florentine *magister*, who taught both in Pisa and Florence and attempted—like his more famous grandson Verino Secondo—to mediate between Platonism and Aristotelianism.[16] In the first lesson, based on *Purg.* 17.91–93 ("Né creator né creatura mai / cominciò ei figliuol, fu sanza amore, / o naturale o d'animo; e tu'l sai") and what follows, Verino closely followed Dante's text, offering a purely theoretical discussion on love. Analyzing Dante's *terzina*, Verino distinguished three kinds of love, the love of the creator, natural love, and the love of the soul, claiming that only this latter can err (*Purg.* 17.94–96); he stated, based on *Purg.* 18.34–39, that not every love, intended as a form, is laudable in itself, even if its matter is always good, and he discussed the definition of love as an inclination toward the thing that is liked and as a spiritual movement toward the thing which is beloved—as in *Purg.* 18.25–27 and 31–33 (*Lettioni* 9). On the basis of these definitions, Verino denied that love as an inclination could belong to the creator, who is never inclined and does not need to be perfectioned by an object of love (*Lettioni* 9). The following questions were trickier: how could natural love always be good, unlike the love of the soul, if the soul is above nature, dealing with spiritual and not bodily objects? The superiority of the soul over nature was also supported by the Platonists, who, according to Verino, had been endorsed by Dante when discussing philosophical matters in his *Convivio*. However, in the case of the *Purgatorio* lines, it is evident that nature—even if inferior to the soul—is closer to God as it acts with no mistakes (*Lettioni* 9). Also, how is it possible that some types of love are not good, when things opposite to love, like hate and disunion, are always negative? How could love not be good if it comes from God? And how could it happen that the matter of love is good when its form is not, as suggested in *Purg.* 18.37–39, even if the opposite would be more comprehensible? And, again, how could Dante define love as an inclination (*Purg.* 18.26), which is a physical movement, not a spiritual one, and how could this passivity be a satisfying definition, when definitions commonly refer to operations? And is it possible to call God love? (*Lettioni* 9–10).

Plato had stated that nature is a cause only as an instrument of the Ideas, and Dante followed him on this point in his *Convivio* claiming that love comes from the Intelligence in charge of the third heaven (*Conv.* 2.3–8) and that in the same way Intelligences would produce all the other generated substantial forms. To explain Dante's statements in *Purgatorio*, Verino embraced instead Aristotle's position, according to which nature is the legitimate cause of things that have bodies or concerning them. This means that human love itself is not caused by the heavens, but by our will and by beauty. After insisting more on the definition of love as a desire for beauty, recovering several images from the *Symposium*, Verino showed that love can be considered both as a movement and as a passion, as Dante says in the *Commedia*, or as a sort of substantial form, a union, as he wrote in the *Convivio*. There is no contradiction in keeping both positions, like Dante did: in the *Convivio* Dante talked as a philosopher, but in the *Commedia*, he preferred speaking of movement, passivity, and desire, all things belonging to the body, because he was talking like a poet in need to be understood by "vulgar men." The love of the *Convivio*, moreover, is a perfected love that does not need anything more, having been fulfilled in the spiritual union with the beloved. However, the love of which Dante wrote in the *Comedy* is still imperfect, as implied by the mention of movement (*Lettioni* 10–12).

Recovering the definition of deliberation from the *Nicomachean Ethics*, Verino showed that the reason why, when speaking of love, matter—that is, the natural desire of the beloved—is always good, but form can be bad, is that form depends on a deliberation that can go wrong. Indeed, certain kinds of love are bad precisely because they are not love but mere shadows of it. Persisting in his approach of interpreting Dante with Dante,[17] Verino found in the *Convivio* the key to solving the issue of how the love of the soul may be subject to error, unlike the inferior natural love. As Dante wrote in *Conv.* 3.3, there are five different loves corresponding to an equal number of natural levels. Human rational nature occupies the highest position in this *scala naturae* and, being more complex, is more easily prone to mistakes. In the same way, the physician masters an art that is more perfect and complex than the art of the weaver and, therefore, will necessarily err more than the weaver (*Lettioni* 13). The question of whether it is possible to call God love finds a double solution: according to the apophatic theology of the Areopagite, when Dante said that God could not be without love (*Purg*. 18.91–92), he meant that God could not be without being the cause of love; according to the doctrine of the prime mover in Aristotle's *Metaphysics* 12, the meaning of the verses would be that by loving Himself, God loves everything—and this everything must be considered not materially, but in the divine mind.

In the second lecture, Verino discussed the incipit of *Paradiso* 1, "La gloria di colui che tutto muove," focusing his attention on love as the mechanism of the world and identifying God with the final cause. In the third lecture, Verino insisted on issues emerging from his previous dissertations, such as the naturality of love and the correspondence between beauty and good. In conclusion, even if his reliance—as a poet—on metaphors and imaginative language in the *Commedia* was not hidden by Verino, Dante emerges from these lectures primarily as a philosopher of love, who had investigated the concept in its profound metaphysical implications, in both Aristotelian and Platonic terms, and not in its daily, mundane occurrences. The communication and complementarity between *Convivio* and *Commedia*, whose doctrines on love contrast in words but not in meaning, not only adumbrates the conciliatory relationship between Plato and Aristotle supported by Verino but confirms Dante's wide-reaching philosophical credentials, which embrace metaphysics as much as natural philosophy.[18]

As we learn from the *Annali dell'Accademia*, in May 1550, Giovanni Cervoni, a collaborator of Bernardo Segni with a preference for Aristotelian philosophy, offered a public reading on *Purgatorio* 18, focused on the concepts of love, "i concetti d'amore."[19] The text was not published and apparently did not survive either in a manuscript form; even if one can speculate that Cervoni embraced more decisively than Verino an Aristotelian perspective, the use of the technical term *concetti* suggests that the content of his lecture was likely not too different from Verino's, in supporting the image of Dante as a philosopher of love.

Instead, we have another lesson on *Purg*. 17.91–93, the same *terzina* at the center of Verino's first lecture. I am referring to the lesson in two parts held by the above-mentioned Benedetto Varchi in 1564, entitled *Sopra que' versi di Dante nel diciasettesimo canto del Purgatorio i quali incominciano: Né creator, né creatura mai, ec.*[20] Despite similarities with Verino's treatment, including the use of the *Convivio* and the interweaving of *Purgatorio* 17 with *Paradiso* 1, Varchi developed his lesson according to a larger scope that embraces all levels of reality. In his argument, the reference to love worked more as a frame that allows him to cover–literally—the entire universe, rather than the subject itself of the discussion. Opening the lesson, Varchi stated:

> That Love that moves the sun and the other stars, love only and only love, my very amorous and loving listeners, was, is and always will be cause not just of the production, but also of the conservation of the entire universe, and of all the things which are contained by it. If love did not exist, therefore, nothing would exist. In fact, the mundane machine would be dissolved and divided, and consequently would collapse; the first mover would not move.[21]

Indeed, he openly affirmed that he did not intend to treat the topic of love "in particolare," having already done that several times, but "generalmente e in universal" (1859: 1.322). Varchi was perfectly aware that in *Purgatorio* 17 Dante focused his discussion on a love that is exquisitely human. Yet, the *Accademico* enlarged his analysis to the entirety of beings, considering the Dantean term *creatura* in its broader sense:

> This word "creatura" means, therefore, anything in a common sense, as long as it is created or produced; hence, it means anything except God. And even if Dante used here this word *creatura* just with reference to rational creatures, that is, human beings, the love at the center of our investigation is instead something that can be found universally in everything, both in the small and big things.[22]

To do so, Varchi implicitly evoked the parallel already suggested by Verino between *Purgatorio* 17 and *Conv.* 3.3, but he complicated this theoretical scheme even more. If Dante had identified in the *Convivio* five types of love corresponding to five natural levels, Varchi delineated a much more elaborated *scala naturae* that highlights a more compact continuity of the universe, close to the model described by Aristotle in his *History of Animals* 8.1. In what follows, Varchi reviewed the ladder of beings, from the prime matter to God. Dante's text, therefore, is bent to the didactic agenda of Varchi: short explanations on the peculiar love characterizing every level of reality merely closed the general discussion on each of them. First matter loves because it desires forms, elements attract and reject each other, and so the imperfect and perfect mixtures, etc. When Varchi discussed human love, he elliptically associated it with free will, faithful to Dante's text, but—more upsettingly—also to a form of intellectual happiness, characterized by a revealing allusion to Averroes's proem to the first book of the *Physics*.[23] The following sections on the heavens, the intelligences, and God implied an Aristotelian cosmos, governed by the prime mover. And again, like in Verino, the treatment of God as the prime mover was modeled according to the first verses of *Par.* 1. Because of the nature of his dissertation, Varchi offered a plurality of definitions of love, and even if love could be said in many ways, he ultimately identified it with a universal force that connects everything according to a different level of complexity. If Verino had eviscerated and reduced to questions Dante's text, Varchi used it to frame a digression on the entire universe. And if Verino used Dante to reconcile Plato and Aristotle, Varchi was more firmly embedded in the Aristotelian tradition, even flirting with Averroes, but still celebrating Plato among his sources. The strategies were different, but the metaphysical merits of Dante's treatment of love were acknowledged in both cases.

Purgatorio 17 was particularly suitable for other kinds of metaphysical readings. In this sense, it is meaningful that different tercets of the canto on the imaginative faculty (13–18) were at the center of a lecture on Dante pronounced in 1589 at the Accademia Fiorentina by Jacopo Mazzoni (1548–1598), a professional philosopher, so confirming the special theoretical appeal of those sections.[24] Yet, it is even more interesting that even if Dante was the most commented author in the public lessons of the Accademia Fiorentina (Gilson 2018: 98), his work was discussed in the context of a lecture on love only in so few instances. Further exceptions, aside from the aforementioned cases, could be Francesco d'Ambra's lessons on *Io son venuto al punto della rota*, in February 1548, and Piero Fabbrini's lecture on *E' m'increscе di me sì malamente*, in June of that same year (Barbi 1890: 92, 228–29). In the absence of clues on the contents of these lectures, the nature of the poems under examination would suggest that both d'Ambra and Fabbrini likely embraced a more "conventional," so to say, approach to love, especially in the case of *E' m'incresce*, that contrasts with the "rational" love treatment that characterized *Purg.* 17–18. If one considers, on the other hand, the overwhelming number of lessons on love based on Petrarch's poetry—both public and private—offered at the Accademia Fiorentina,[25] it becomes evident that the *Accademici* accepted Petrarch's expertise on love, but also agreed that the scope of their discussions on the same topic had to invest a superior theoretical outlook when Dante was invoked. Dante was at the basis of dissertations dealing with love in definitory terms and according to an exquisitely metaphysical dimension of universal love. Petrarch was the privileged interlocutor when dealing more concretely with human love. Petrarch could give support to understand why one blushes or gets pale in the presence of the beloved, but Dante showed what keeps the universe together. Read as Platonic as much as Aristotelian, Dante allowed the *Accademici* to identify a space in which love can be discussed with no use for the affections of mind and body, painstakingly analyzed and explored in Petrarch's lyrics. And, certainly, despite all philosophical labels, the long shadow of Ficino, one of the tutelary deities of the *Accademia Fiorentina*, was not alien to this attitude.

As once again Varchi wrote, Cavalcanti and Dante spoke of love in a way that was different from the one employed by other poets, not simply of its accidents and effects, but of its substance and nature: they spoke poetically, but as philosophers.[26] In fact, in the *Accademici*'s lectures on Dante and love we have discussed, Dante appeared simply as a philosopher, with no qualifications of any sort. It is therefore notable that in sixteenth-century commentaries on Guido's *Donna me prega*, Dante's friend was instead described either as a "filosofo naturale" (as already in *Decameron* 6.9) and occasionally as "filosofo morale" (Alfano 2010).[27] Guido, after all, focused on destructive, material love as the main object of his investigation. But Dante's treatment of love—that connects the physical world to the divine, bridging them with ethics via the crucial issue of free will—could not be characterized univocally since it embraced all philosophical branches. In a subject like love the tension between *res* and *verba*, magisterially studied by Simon Gilson in the context of Dante's sixteenth-century reception, was—in conclusion—particularly extreme.

I started this chapter by pointing out the absence, or at least the limited presence, of Dante in Renaissance love treatises. The lectures at the Accademia Fiorentina are especially notable and revealing in this context. They show that Dante could be the poet-philosopher of love without being second to Petrarch once the subject was to be discussed according to an abstract and metaphysical frame. However, something is still missing: despite this vision of love—emphasizing, for example, the importance of will—could appear more reassuring in the counterreformation period, when love treatises displayed more openly theological tones (Del Soldato forthcoming), Dante's presence in them remained secondary.[28] Dante, the philosopher of love, was at that point likely a casualty of the general decline of interest toward him as a poet. Still, at least in a precise context, the Accademia Fiorentina, he had been able to make his voice heard when speaking of love. Closing again with Sherlock Holmes, that dog, after all, barked.

Notes

1. Giuseppe Zonta edited a selection of these treatises (*Trattati* 1912). See also Nelson (1958: 67–162).
2. See Fenzi (1999). There is of course an extensive bibliography on Cavalcanti in the medical and philosophical traditions. See, among the most recent contributions, Pace (2016). On Cavalcanti's early "authoritativeness" among physicians and philosophers, see Barański (2003).
3. On their struggle see Aasdalen (2011).
4. Niccoli (1987). On the history of the text, see Roggia (2019).
5. See Tonelli (2021: 232). Here and below, translations are mine if not otherwise noted.
6. Indeed the only other poem in which Beatrice's name is explicitly mentioned, but in which she is associated, *à la* Cavalcanti, to Dante's death and destruction (*Lo doloroso amor che mi conduce*) was not included in the *Vita nuova*. See on this Pazzaglia (1970–1978) and Barolini (2006: 70–101).
7. See, at least, Barański (2001: 153–73; 2007); Gilson (2009: 38–39). On Machiavelli see the letter to Francesco Vettori, December 10, 1513, Gilson (2018: 288). For Savonarola's hostility against Dante as a vernacular love poet, see Gilson (2018: 35). On the "rivalry" between Dante and Petrarch as lyric authors in the fifteenth and sixteenth centuries, see Banella (2020) and Tomasi (2021).
8. Lenzoni (1556: 202): "Tratta di poche cose oltra lo amore" vs. "Tratta di infinite." On the work Gilson (2018: 110–15). Already in the *Convivio* Dante presented himself as an "expert not just in matters of love, but also in a wide range of other doctrinal issues" in opposition with Cavalcanti, see on this Barański (2004:

20). Gelli (1551: 57) ranked Petrarch as a philosopher after Dante, even when praising his profound sonnets of repentance, opposing them to the previous ones, which "non trattano che di passioni amorose."

9. Varchi (1859: 2.527): "E se bene io conosco dove entro e qual rischio mi metto, non perciò voglio tacere quello di che molti dubitano e tenzonano, ciò è chi abbia meglio d'amore scritto o Dante o il Petrarca, dicendo che se per quella parola *meglio* intendono più dottamente e più veramente e in somma più da filosofo, io per me non dubito che Dante n'abbia il meglio; ma se per *meglio* intendono o più o più leggiadramente, senza dubbio il Petrarca, a mio giudizio, starà sopra...". Giovanni Pico had suggested the same opposition between form and content in Dante and Petrarch (Bausi, 1998: 24, 26, 28).
10. On Dante as part of the medieval tradition of the poet-philosophers, and his innovations—the abandonment of "veils" in favor of a treatise-like expression—see Boyde (1981).
11. Tomitano (1546: 96). On this text, see Barbi (1890: 17) and Gilson (2018: 111). Yet Tomitano (1546: 44) praised Petrarch also as a philosopher.
12. An exception in Lenzoni (1556: 122–23), in which he quotes *Inf.* 5.100–06 and 127–42.
13. For more occurrences of *Inf.* 5.103, see Favaro (2012: 170–75).
14. But see again Lenzoni (1556: 118–19), in a section expressly devoted at showing that Dante, especially in the *Commedia,* "was no less extraordinary and excellent in [describing] love's effects than in any other thing he treated." According to Giovambattista Gelli, Dante wrote on several subjects—including love—way better than classical predecessors (in the case of love those were Catullus, Propertius and Tibullus). See Gelli (1887: 16). For a fifteenth-century use of *Purg.* 17, aimed at presenting Dante as an authority on love, see Banella (2020: 178).
15. But see Petterutti Pellegrino (1998: 198), who shows Equicola's dependence on Landino's commentary.
16. The lectures are published in *Lettioni d'Academici Fiorentini sopra Dante, libro primo* (Florence: [Doni], 1547), 9–24, hereafter *Lettioni*. See also Gilson (2018: 99). On Verino see Carotti (2020).

17. On commenting Dante with Dante see Gilson (2018: 86, 103, 134, 159, 161, 180, 247, 310). On this exegetical practice, common among Aristotelians, see Bianchi (1996).
18. On the agreement between Plato and Aristotle see Del Soldato (2020).
19. Barbi (1890: 230). According to the *Annali*, (Florence, Biblioteca Marucelliana, MS B III 54, fol. 5v), Cervoni became accademico in 1542. On Cervoni see also Bionda (2002: 421–23); Wainwright (2017: 2).
20. On Varchi and Dante, see at least Gilson (2018: 115–25).
21. Varchi (1859: 1.322): "l'amor, che muove il sole e l'altre stelle, l'amor solo, solo l'amore, amorosissimi e amorevolissimi ascoltatori, fu ed è e sarà sempre cagione non solamente della produzione, ma eziandio della conservazione di tutto l'universo mondo, e di tutte le cose che in tutto l'universo mondo si contengono. Conciossia cosa che se non fusse amore, non sarebbe cosa nessuna: perché, oltra che si dissolverebbe e disunirebbe, e per conseguente mancherebbe tutta questa macchina mondana, il primo motore non moverebbe."
22. Varchi (1859: 1.323): "Significa dunque questo verbale creatura comunemente qualunque cosa, la quale sia creata o prodotta, e per conseguente ogni cosa sia qual si voglia, eccetto Dio. E se bene Dante la ristrinse qui alle creature razionali sole, cioè agli uomini, non è che l'amore, del quale intendiamo di favellare, non si ritruovi universalmente in tutte le cose, dalle picciole alle grandi."
23. Varchi (1859: 1.329): "Le quali tutte cose sono state dette da me così brevemente, perché dimostrano tutte insieme, e ciascuna di loro, che nella spezie umana si truovi il maggiore e il migliore amore che trovare si possa in cosa alcuna sotto il cielo. Perché l'uomo solo, tra tutte le creature o mondane o celesti ha l'arbitrio libero, ed è capevole delle virtù e delle scienze: mediante le quali egli può (come io ho detto pur testé) copulare l'intelletto possibile e l'agente, e divenire quasi Dio. E per questo diceva quel medesimo Arabo, che uno uomo che sappia, e uno che non sappia sono equivoci, cioè si possono bene chiamare uomini, ma non sono: perché hanno solamente il nome comune, ma non la sostanza. E il gran Filosofo diceva nella Politica, che come l'uomo buono è il migliore animale che sia, così il reo è il più cattivo: e l'essere l'uomo o buono o reo viene (come tutti gli altri o beni

o mali) solamente dall'amore." On this passage see also Bianchi (1992: 197). On Averroes's prolog see Retucci (2021), which also discusses it as possible source to Dante's *Convivio*, as suggested by Bianchi (2018).

24. Barbi (1890: 191, 235). Mazzoni wrote a *Difesa della Commedia di Dante* (Cesena: Raverii, 1587), in which he also discussed *Purg.* 18.49–66 to explain what *phantasia* is.
25. See the *Annali* preserved in Florence, Biblioteca Marucelliana, MSS B III 52–54.
26. Varchi (1859: 2.527): "E a chi dicesse che non pur Guido Cavalcanti, nella sua oscurissima canzone, ma ancora Dante ne favellarono diversamente degli altri, raccontando non gli accidenti ed effetti di lui solamente, ma ancora la natura e sostanza sua, direi che ne favellarono poeticamente e non come poeti, ma come filosofi." On the early perceptions of Guido as more a philosopher than a poet—also in the ancient commentaries to the *Commedia*—see Inglese (1992: 86–89).
27. Ficino (1987: 177–79) defined Guido as "philosopher" in the Socratic (hence Platonic) tradition. Medieval interpreters like Benvenuto da Imola defined Dante as a poet and Guido as a philosopher (1887: 1.341) to stress their complementarity. See also Barański (2003). On Dante rejecting Guido's authority on love see Barolini (2006: 70–101).
28. One could think about the crucial issue of free will, on which see Favaro (2012: 179–82).

References

Aasdalen, Unn Irene. 2011. The First Pico–Ficino Controversy. In *Laus philosophi platonici. Marsilio Ficino and His Influence*, eds. Stephen Clucas, Peter J. Forshaw, and Valery Rees, 67–88. Leiden: Brill.

Alfano, Giancarlo. 2010. 'Guido filosofo'. I commenti cinquecenteschi a *Donna me prega* nel loro contesto culturale. *Filologia e Critica* 25: 3–43.

Alighieri, Dante. 1988. *Il convivio*, eds. Cesare Vasoli and Domenico De Robertis, 2 vols. Milan-Naples: Ricciardi.

Arnigio, Bartolomeo. 1566. *Dialogo della medicina d'amore*. Brescia: Marchetti.

Banella, Laura. 2020. *Rime e libri delle rime di Dante tra Medioevo e primo Rinascimento*. Rome: Edizioni di storia e letteratura.

Barański, Zygmunt G. 2001. *"Chiosar con altro testo." Leggere Dante nel Trecento.* Fiesole: Cadmo.

Barański, Zygmunt G. 2003. Guido Cavalcanti and His First Readers. In *Guido Cavalcanti tra i suoi lettori*, ed. Maria Luisa Ardizzone, 149–75. Fiesole: Cadmo.

Barański, Zygmunt G. 2004. "Per similitudine di abito scientifico": Dante, Cavalcanti and the Sources of Medieval "Philosophical" Poetry. In *Science and Literature in Italian Culture: from Dante to Calvino. A Festschrift for Patrick Boyde*, eds. Pierpaolo Antonello and Simon Gilson, 14–52. Oxford: Legenda.

Barański, Zygmunt G. 2007. Petrarca, Dante, Cavalcanti: La formazione dell'*auctoritas* volgare. *Deutsches Dante Jahrbuch* 82: 119–146.

Barbi, Michele. 1890. *Dante nel cinquecento.* Pisa: Nistri.

Barolini, Teodolinda. 2006. *Dante and the Origins of Italian Literary Culture.* New York: Fordham University Press.

Bausi, Francesco. 1998. L'epistola di Giovanni Pico della Mirandola a Lorenzo de' Medici: Testo, traduzione e commento. *Interpres* 17: 7–57.

Benvenuto da Imola. 1887. *Comentum super Dantis Aldigherij Comoediam*, ed. Giacomo Filippo Lacaita, 5 vols. Florence: Barbera.

Bianchi, Luca. 1992. Filosofi, uomini e bruti, note per la storia di un'antropologia averroista. *Rinascimento* 32: 185–201.

Bianchi, Luca. 1996. 'Interpretare Aristotele con Aristotele': Percorsi dell'ermeneutica filosofica nel Rinascimento. *Rivista Critica Di Storia Della Filosofia* 51: 5–27.

Bianchi, Luca. 2018. 'Ultima perfezione' e 'ultima felicitade.' Ancora su Dante e l'averroismo. In *Edizioni, traduzioni e tradizioni filosofiche (secoli XII–XVI). Studi per Pietro B. Rossi*, 2 vols., edited by Luca Bianchi, Onorato Grassi and Cecilia Panti, vol. 1, 315–28. Rome: Aracne.

Bionda, Simone. 2002. La copia di tipografia del *Trattato dei Governi* di Bernardo Segni: Breve incursione nel laboratorio del volgarizzatore di Aristotele. *Rinascimento* 42: 409–442.

Boyde, Patrick. 1981. *Dante Philomythes and Philosopher: Man in the Cosmos.* Cambridge: Cambridge University Press.

Carotti, Laura. 2020. Vieri, Francesco de', detto Verino primo. *Dizionario biografico degli italiani*, vol. 99. https://www.treccani.it/enciclopedia/vieri-francesco-de-detto-verino-primo_%28Dizionario-Biografico%29/. Accessed January 18, 2023.

Del Soldato, Eva. 2020. *Early Modern Aristotle. On the Making and Unmaking of Authority.* Philadelphia: University of Pennsylvania Press.

Del Soldato, Eva. Forthcoming. Crivellati e gli altri: la malattia di amore nella Controriforma. In *Medicina, magia, miracoli a Roma nella prima età moderna*, eds. Gerardo De Simone and Francesca Niutta. Rome: Roma nel Rinascimento.

Dolce, Lodovico. 1560. *Dialogo della institution delle donne*. Venice: Giolito.
Ebbersmeyer, Sabrina. 2002. *Sinnlichkeit und Vernunft. Studien zur Rezeption und Transformation der Liebestheorie Platons in der Renaissance*. München: Verlag.
Equicola, Mario. 1526. *Libro di natura di amore*. Venice: Nicolini da Sabbio.
Favaro, Maiko. 2012. *L'ospite preziosa. Presenza della lirica nei trattati d'amore del '500 e del primo '600*. Lucca: Pacini Fazzi.
Favaro, Maiko. 2021. *Ambiguità del petrarchismo. Un percorso fra trattati d'amore, lettere e templi di rime*. Milan: Franco Angeli.
Fenzi, Enrico. 1999. *La canzone d'amore di Guido Cavalcanti e i suoi antichi commenti*. Genoa: Il melangolo.
Ficino, Marsilio. 1987. *El libro dell'amore*, ed. Sandra Niccoli. Florence: Olschki.
Gelli, Giovambattista. 1551. *Tutte le lettioni*. Florence: Torrentino.
Gelli, Giovambattista. 1887. *Letture edite e inedite sopra la* Commedia *di Dante*, ed. Carlo Negroni, 2 vols. Florence: Bocca.
Gilson, Simon. 2009. *Dante in Renaissance Florence*. Cambridge: Cambridge University Press.
Gilson, Simon. 2018. *Reading Dante in Renaissance Italy. Florence, Venice and the "Divine Poet."* Cambridge: Cambridge University Press.
Inglese, Giorgio. 1992. Per Guido "filosofo" (*Decameron* VI, 9). *La cultura* 30: 75–95.
Lenzoni, Carlo. 1556. *In difesa della lingua Fiorentina e di Dante, con le regole da far bella et numerosa la prosa*. Florence: Torrentino.
Lettioni d'Academici Fiorentini sopra Dante, libro primo. 1547. Florence: [Doni].
Monfasani, John. 1993. Aristotelians, Platonists, and the Missing Ockhamists: Philosophical Liberty in Pre-Reformation Italy. *Renaissance Quarterly* 46: 247–276.
Moyer, Ann E. 2020. *The Intellectual World of Sixteenth-Century Florence. Humanists and Culture in the Age of Cosimo I*. Cambridge: Cambridge University Press.
Nelson, John Charles. 1958. *The Renaissance Theory of Love: The Context of Giordano Bruno's Eroici Furori*. New York: Columbia University Press.
Nobili, Flaminio. 1590. *Trattato dell'amore humano*. Bologna: Bonardo.
Pace, Matteo. 2016. L'amore di Cimone. Tradizione medica e memoria cavalcantiana in *Decameron* V 1. *Studi Sul Boccaccio* 44: 251–275.
Pazzaglia, Mario. 1970–1978. Lo doloroso amor che mi conduce. In *Enciclopedia Dantesca*, ed. Umberto Bosco. Rome: Istituto dell'enciclopedia italiana. https://www.treccani.it/enciclopedia/lo-doloroso-amor-che-mi-conduce_%28Enciclopedia-Dantesca%29/. Accessed January 14, 2023.
Petterutti Pellegrino, Pietro. 1998. Tessere landiniane nelle *Institutioni* dell'Equicola. Annotazioni a margine di un caso di intertestualità dissimulata, ovvero

di un accorto e misurato plagio. In *Furto e plagio nella letteratura del Classicismo*, ed. Roberto Gigliucci, 161–217. Rome: Bulzoni.

Plaisance, Michel. 2004. *L'accademia e il suo principe. Cultura e politica a Firenze al tempo di Cosimo I e di Francesco de' Medici*. Manziana: Vecchiarelli.

Retucci, Fiorella. 2021. Due prologhi di Averroè alla *Fisica* e la 'ultima felicitade' umana. *Giornale Critico Di Storia Della Filosofia* 100: 457–472.

Roggia, Enrico. 2019. Ficino's Self-Translation of the *De amore*: Some Linguistic Remarks. *Rivista Di Storia Della Filosofia* 30: 193–211.

Speroni, Sperone. 1596. *Dialoghi*. Venice: Meietti.

Tomitano, Bernardino. 1546. *Ragionamenti della lingua toscana*. Venice: Farri.

Tomasi, Franco. 2021. Dante e i poeti "moderni" nel primo Cinquecento: primi appunti. In *Oltre la Commedia. Dante e il canone antico della lirica (1450–1600)*, eds. Laura Banella and Franco Tomasi, 41–57. Rome: Carocci.

Tonelli, Natascia. 2015. *Fisiologia della passione. Poesia d'amore e medicina da Cavalcanti a Boccaccio*. Florence: Edizioni del Galluzzo.

Tonelli, Natascia. 2021. Medicine. In *The Oxford Handbook of Dante*, ed. Manuele Gragnolati, Elena Lombardi, and Francesca Southerden, 227–241. Oxford: Oxford University Press.

Trattati d'amore del Cinquecento, ed. Giuseppe Zonta. 1912. Bari: Laterza.

Varchi, Benedetto. 1859. *Opere... ora per la prima volta raccolte*. 2 vols. Trieste: Lloyd.

Wainwright, Anna. 2017. A Simple Virgin Speaks: Authorial Identity and Persuasion in Isabella Cervoni's *Oration* to Pope Clement VIII. *The Italianist* 37: 1–19.

Index

M. Pace (ed.), *Dante and the Sciences of the Human*, Palgrave Studies in Medieval and Early Modern Medicine,
https://doi.org/10.1007/978-3-031-69253-6

SPRINGER NATURE

GPSR Compliance

The European Union's (EU) General Product Safety Regulation (GPSR) is a set of rules that requires consumer products to be safe and our obligations to ensure this.

If you have any concerns about our products, you can contact us on ProductSafety@springernature.com

In case Publisher is established outside the EU, the EU authorized representative is:

Springer Nature Customer Service Center GmbH
Europaplatz 3
69115 Heidelberg, Germany

The manufacturer's authorised representative in the EU is Springer
Nature Customer Service Centre GmbH, Europaplatz 3, 69115 Heidelberg,
Germany. If you have any concerns regarding our products, please
contact ProductSafety@springernature.com

Printed and bound by CPI Group (UK) Ltd, Croydon, CR0 4YY
30/09/2026
02225216-0002